CURING THE INCURABLE

New Techniques and Miraculous Cures

by

JOSHUA D. SALVADOR, M.D. FRCS (C)

"Look outside the box! Innovate!"

**FOREWORD BY
DENTON A. COOLEY, M.D., FACS.
SURGEON-IN-CHIEF & FOUNDER OF
THE TEXAS HEART INSTITUTE AND
THE COOLEY HEART INSTITUTE**

ISBN 978-1-304-19520-3

Edited by Jeffrey Beal.
Interior Design and Layout by James L. King III.
Cover by Gregory Clark of Tek Ninja Design.

Printed in the United States of America

Typeset using Scribus 1.4, using the typefaces Volkorn and Source Sans Pro

Published September 2013

This book is dedicated to

Denton A. Cooley, M.D.

A unique and compassionate mentor,
and a great heart surgeon and innovator.

"Most of the impossible things in the world have been accomplished by people who kept on trying where there seemed to be no hope."

—Dale Carnegie

"It is difficult to say what is impossible, for the dream of yesterday is the hope of today and the reality of tomorrow."

—Robert Goddard

"Hope is the conviction that something will turn out well regardless of how it turns out."

—Author Unknown

"Hope sees the invisible, feels the intangible, and achieves the impossible."

—Author Unknown

"Those of us who hope strongest have within us, the gift of miracles."

—Sidney Bremer

"With a single good deed, you will tip the scales for yourself and the entire world to the side of good."

—Moses Maimonides

If I can stop one heart from breaking,

I shall not live in vain;

If I can ease one life the aching,

Or cool one pain,

Or help one fainting robin

Unto his nest again,

I shall not live in vain.

—Emily Dickinson

TABLE OF CONTENTS

FOREWORD

by Denton A. Cooley, M.D.,
Surgeon in Chief of the Texas Heart Institute

This collection of Miraculous Cures by Josh Salvador reveals an unusual curiosity and inquisitiveness of a surgeon for outcomes in the treatment of patients who seemed to be beyond cure. In searching for answers, he explored the values of some additional factors such as intuition, innovation, compassion, prayer, faith, creativity, attitude, serenity, the mind-body connection and some alternative medicines. All these factors together complement traditional medicine and resulted in the remarkable cures, which he considers "miraculous."

Having known Josh since his training years as a fellow in my service at the Texas Heart Institute, I appreciate his devotion to the patient's welfare and to those teachers and role models whose contributions influenced his career.

One must appreciate that his compassion for them enhanced their cure. Since Salvador established a compassionate and friendly relationship with them, which is sometimes lacking in routine medical practice, such devotion to their health care made them wish to give him satisfying results that enhanced their cure. Sometimes this could be a beneficial placebo effect.

Of interest to me is that Dr. Salvador, a retired surgeon, collected stories from some of his previous patients, explained the modifications of the usual techniques and alternatives for his incurable patients that resulted in their remarkable recovery. These actions made his experience as a physician rewarding and fascinating.

For the young person who is considering medicine, the message is clear that it is an attractive, interesting and satisfying career if he is willing to undertake the often difficult academic process involved, and add his own contribution to it.

This book is of interest to both laymen and physicians. Although I am not inclined to endorse all of the author's observations at this time, I do adhere to his advice that the physician should maintain an open mind to innovations, look for new and evolving techniques, and exclude those that he considers have outlived their values.

—Denton A. Cooley, M.D.

INTRODUCTION

"If you save one life, you have saved the whole world."

—Hebrew Proverb

I offer this book for the benefit of patients, surgeons, medical practitioners, medical students and the medical profession as well. It's the result of fifty years of evolving medical care, and from the deepest part of my being.

In spite of the great discoveries in medicine, that profession has its own limitations. Until now there are diseases which have been considered incurable or statistically fatal. So, medicine, as yet, is not a perfect science, but one that is seeking perfection.

During my career I've found that depending on standard medical protocols can lead to incorrect diagnoses, wrong treatments and erroneous predictions which are all too often unnecessarily gloomy. So, I didn't accept everything written in medical literature as unerring fact, and started to think differently. When I could not find a cure for a patient, I looked beyond the protocols to alternative methods. I also used my intuition, creativity and imagination. This led me to be inspired to methods which saved lives and diminished suffering, replacing the old ways which led to the negative prognosis. I used these new techniques over and over again on other patients who were also considered hopeless or incurable. They proved to be invaluable, and I hope they'll be of value to other surgeons, and inspire them to do likewise. After all, that's how medicine progresses. What was incurable yesterday becomes curable today. To make a difference in other people's lives is one of the greatest rewards for a surgeon. For all those reasons, "Looking outside the box," is the theme of this book.

The book format is in two parts:

PART I

In the first part I've chosen the stories of several patients to exemplify each new technique, though this limited space doesn't permit mention of all who've benefited. What's important to remember is that when something's considered beyond salvage or cure, think, *"There may be another way!"* This includes patients who were saved from fatal cancers, whether of the lung or ovaries, malignant lymphomas or metastatic cancer. In breast cancer I challenged the postoperative protocol of the routine use of irradiation and chemo after a complete lumpectomy which can cause more harm than good. Other patients were saved from amputations, or cured from vascular lesions considered beyond help. Some needed emergency heart surgery, while others were in prolonged coma, thought to be beyond recovery. Dangerous operations to relieve nerve compression were simplified and made safer by new modifications, while still others suffered fatal traumas. The value of a positive attitude, along with hope and optimism on the part of the patient and doctor, cannot be over emphasized, and is demonstrated herein with many examples.

PART II

In this part you'll find the attributes needed by both patients and doctors to achieve remarkable salvations. Many intriguing subjects are also discussed, such as the power of prayer and meditation, miraculous spontaneous cures, regression of tumors, spiritual or faith healing, and whether, by coincidence or destiny, I was led to the scene of a dying patient without planning to be there.

The last chapter in this part concerns the treatment of war casualties, particularly those resulting from conflicts between Egypt and Israel. This added tremendously to my experience in saving fatally injured patients in civilian life. They're also interesting from a historic point of view.

To better clarify some medical conditions for the average reader, illustrations have been added. "One picture is worth a thousand words."

Much of the information in this book cannot be found in any other book, and thus will be of great value to patients and doctors alike.

Doctor/patient relationships are very important for cures. At present it suffers from over-mechanization. There's far too much dependence on new diagnostic tools like CT scans, MRI's, PET scans, blood tests, etc. Patients are demanding from doctors the return to personalized care, listening attentively to their complaints, performing accurate physical examinations, and attending to their emotional and spiritual needs.

Most medical schools teach their students medicine as a science, and to use it as taught. They don't teach them to think beyond the books, nor how to interact with their patients. It's my hope this book will fill that need. And to the patients who read this book, I urge you not to despair or lose hope. Just because there was no cure yesterday, doesn't mean there won't be one tomorrow. Medical science is constantly progressing and discovering new cures.

ACKNOWLEDGMENTS

It's hard to express my heartfelt gratitude to those who helped, inspired and motivated me in writing this book.

First, to my patients who wanted to share their stories with others who might find themselves afflicted with similar "hopeless" situations, their message is, *"Never lose hope; there might be a way for salvation."*

To my wife, Patricia, RN, whose encouragement helped me overcome difficulties that came along. She's a rock I can lean on in hard times, and my pillar of support in health and sickness, in success and shortcomings. She shared my compassion for the patients whose stories you are about to read.

To my secretary, Alma, who collected all the data regarding the patients, and typed, edited, and classified them into categories according to their type of disease, I'm most appreciative.

To Rodelia, my office manager of twenty years who knew every patient in this book, often counseling and encouraging them.

To Mr. Hector Duarte, for reviewing this book, my deepest gratitude.

To my medical illustrator, Sarita Hleap, I cannot fully express my gratitude and appreciation for her artistic talent, motivation and encouragement.

I thank Alexandra Kronfeld for her artistic advice in the design of this book.

I'm eternally grateful to my parents for emphasizing the value of education, for giving me the opportunity to study and be what I am, and for supporting me in the pursuit of my ambitions.

I was blessed to have great mentors who trained and inspired me, most of all, Dr. Denton Cooley. I revere his spirit of innovation, creativity, dedication to perfection, modesty, sense of humor and serenity during surgery. I also admire him for keeping his dignity when facing criticism, as when he dared to put an artificial heart in a patient for the first time and prove that a human being can remain alive with it. In so doing, he opened the way for research in this direction. Without doubt, he's my idol and role model.

I'm grateful for the opportunity to have worked with Dr. Albert Starr, the first surgeon to put a metal valve in a human heart, disregarding criticism or consequences. His pioneering spirit, ingenuity and courage are all qualities that have influenced me and enhanced my career.

I'd also like to acknowledge the pioneering spirit of Dr. Pawzner, who, in the 1960s, developed innovations in lung surgery and closed heart surgery. He also initiated the first open heart program in Israel.

To Dr. M. Deanzman, who opened the first pediatric surgery unit in Beilinson Hospital at Tel Aviv University where I had most of my general surgical training. I have great regard for his elegance and surgical dexterity.

I also remember all of my tutors at the University of Alexandria Medical School. They were all Fellows of the Royal College of Surgeons of England, and taught me the early principles of surgery. To this day, over fifty years later, I remember all of them by name.

In addition, I value all those who guided my internship. I cannot forget Professor Julian Taylor, who was the Vice President of the Royal College of Surgeons of England, and Mr. McGowan, who later, in 1965, became the President of the Royal College of Surgeons of Ireland.

I'm also appreciative to Mrs. Samson Wright who found me a job in England when I was in desperate need of one. Her husband, Sir Samson Wright, wrote the classic book, *Applied Physiology.*

I cherish the memory of every tutor with gratitude and respect.

Finally, I wish to acknowledge Rob Eastaway, author of the book *Out of the Box*, for inspiring so many of us to think creatively.

To any others I may have forgotten my appreciation is still most sincere. I thank you all.

Last but not least, I acknowledge with deep appreciation the efforts of James L. King III to reconstruct the format of the book, making it easier to read and comprehend.

DISCLAIMER

This book is about saving lives in situations considered hopeless or with little chance of cure. The author believes there are no incurable diseases, but diseases whose cures are not yet discovered. Towards that goal, we in the medical profession should walk the extra mile and ask ourselves, "Is there something more that can be done?" This may lead to discovering new ways to treat diseases. To do so requires love and compassion for our patients, and honoring the doctor-patient relationship.

Much of the information and techniques in this book are new and have not been published in other medical literature. They can be used to obtain positive results when applicable. However, the author will not take responsibility for any complications resulting from their misuse.

Other doctors and surgeons should look outside the box, be creative, ask for guidance and inspiration, be humble and open minded to new alternatives when the usual and customary techniques are not working, and discover their own new ways for curing the incurable.

Names of patients in this book have been changed to protect their privacy, except where confidentiality was waived.

PART I

PATIENT STORIES AND COMMENTARIES

SECTION I

CANCER CURED

CANCER

"Be still, Let go. The battle is God's, not yours. And because it is God's battle through you, victory was on your side before the battle began."
—Emilie Cady

INTRODUCTION

The fear of cancer kills more people than cancer itself. When faced with this crisis, it's important to remember not to panic. Many cancers are curable. For example, breast cancer, when detected early, can be cured with a simple lumpectomy.* Also, there are cancers of the skin that can be cured by mere excisions. Other cancers are slow growing, and people can live with them for many years if they are not frightened. Examples of these include thyroid and prostate cancer. In addition to that, cancer can undergo *spontaneous regression*. Most important of all is to not be terrified of the "Big C" as if it were the Bogey Man, dreaded by everyone. Unreasonable fear will bring about your demise quicker than any illness, including cancer, and your positive frame of mind will contribute to your wellness more than you know. That's one of the principal messages of this book. In these pages you'll learn about audacious medical techniques, from drugs to surgery, and the people they saved. Running through these stories is the recurring theme that an affirmative attitude is an effective and miraculous cure. It's not sentimentality, or a pathetic myth to latch onto because

*—*Please refer to the glossary in the back for medical terms.*

5

there is no other hope. Not at all. A positive mindset, on the part of the doctor, as well as the patient, can make all the difference. As you'll see in these histories, a positive thinking doctor can find a cure where status quo physicians can't. It's often just practical. Seek and you shall find. All too often, doctors become fatalistic regarding cancer, assuming that fate will take over and their patients will succumb. They don't even bother to seek a solution. But every person is an individual with a distinct body, and they react differently to similar stimuli. A doctor who believes this will search out the particular remedy for that unique individual. It's just common sense, and not a logical example of the power of positive thinking. This book has many more. Read on.

Cancer research has been progressing at a higher rate than ever before. Some brain cancers can be treated by tumor embolization, whereby the blood supply of the tumor is occluded by administrating materials to block the arteries that feed them using a catheter introduced through the groin, eliminating the need to open the skull.

Inspired by love for my patients, I have developed methods of surgically treating metastatic tumors and lymphomas. The dogma that they are not surgical targets has been refuted. In breast cancer, the use of postoperative irradiation and chemo (the protocol) has been challenged.

Research is currently being conducted on removing the cancer genes when they are hereditary, as is the case with many cancers. The mechanism of accelerated cell division and how to control it is also being actively researched. Knowledge of the mechanism of cell division could help. Cell division occurs in our bodies throughout our lifetime whereby every cell in every organ divides to replace the old or dead cells. It is controlled by a type of meter (the sarcomere) which determines the number of times a cell can divide. Cancer occurs when this mechanism gets out of control. I feel it won't be long before the enzyme that shortens the sarcomere of cancer cells will be discovered, which will stop their cell division and ultimately the cancer growth.

In the meantime, the greatest hope for cure depends on early detection, whereby the cancer can be removed before it spreads. In women, breast cancer can be discovered by self-examination, routine mammograms or ultrasounds. And cancer of the cervix can be discovered by routine pap smears.

In men, cancer of the prostate can be detected early by routine blood test for PSA, cancer of the colon by routine colonoscopies, and cancers in hidden places by blood CEA (cancer embryonic antigen).

If any of these test results come back positive, don't panic. Your initial anxiety is expected, but it must be followed by hopefulness, secure in the knowledge that by early detection your cancer can be removed by the proper specialist. Choose a doctor you're comfortable with, and avoid those who are clumsy, unsympathetic and pessimistic. Pray for guidance to lead you to the proper doctor.

The following stories bear witness to the fact that any cancer is most certainly treatable.

In the coming pages, we will discuss lung cancer, breast cancer, metastatic cancer, and lymphomas (cancer of the blood).

LUNG CANCERS

INTRODUCTION

Lung cancer usually occurs in heavy smokers and develops about or after the age of 40; most of the time it is asymptomatic, detected by a routine chest x-ray.

Cancer of the lung can be primary, originating in the lung itself, or metastatic, which means it spread to the lung from cancer in another organ. These are usually peripheral, (occurring close to the surface of the lung) while primary cancer can be central, meaning close to the heart.

There are different types of cancer depending on the type of cell in the lung that produces them: squamous carcinomas (35%), adeno carcinomas (25%), undifferentiated carcinomas (15%), small cell carcinomas (20%) and alveolar cell carcinomas (5%). The worst is the small cell variety, whereby the cells divide and multiply before they even mature and travel rapidly to the blood stream before they are detected. That's why they're considered inoperable and usually not referred for surgery, but treated with chemo. The average longevity after discovery and chemo is two years. Most of the other varieties are resectable (able to be surgically excised) if they have not metastasized to other organs by the time of detection. That is why early detection is of great value, especially in the case of smokers by a routine x-ray, sputum examinations or CEA blood tests.

A small percentage of lung tumors occur inside the lumen of the bronchial tree. These tumors, whether benign or malignant, are big trouble makers and should be resected as soon as discovered. They could obstruct the air entry to part of the lung, causing mucous secretions to accumulate and get infected, or they can bleed and drown the patient in his own blood. The first type causes cough and fever; the second causes hemoptysis (coughing of blood).

Figure I will show you the anatomy of the lung and its function. Unoxygenated blood comes to the lung by the pulmonary artery and after being oxygenated, goes back by the pulmonary veins to the left ventricle, pushing it through the aorta to be distributed to all parts of the body.

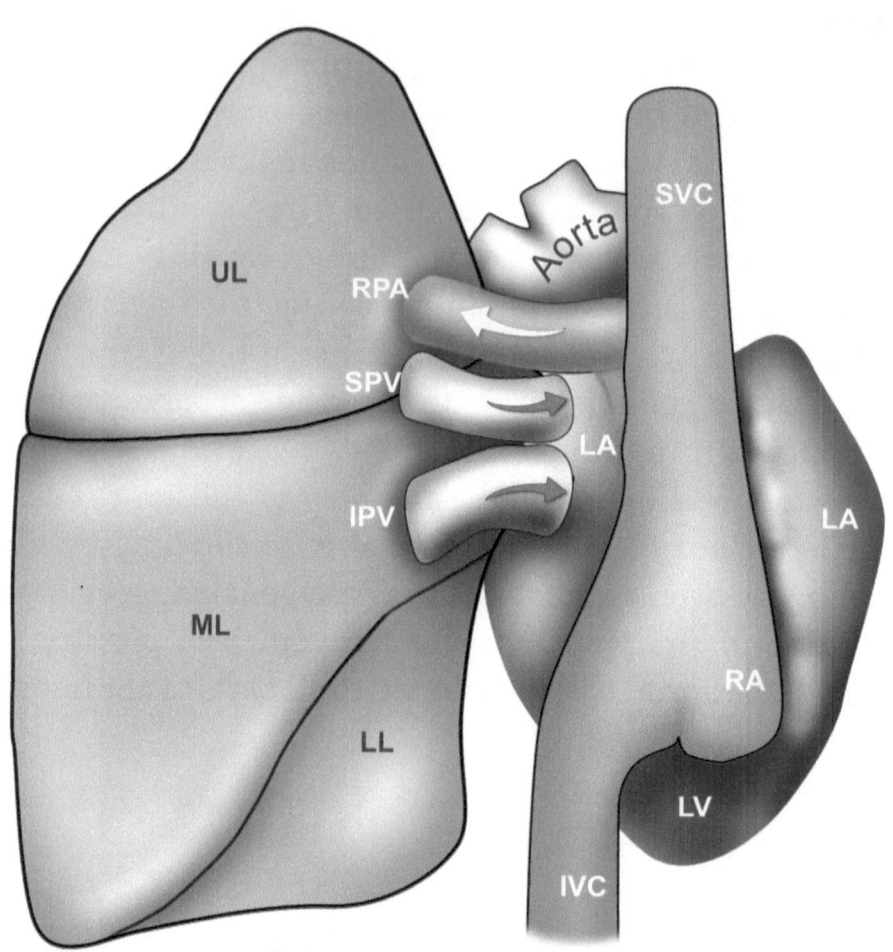

THE CIRCULATION OF BLOOD BETWEEN HEART AND LUNGS:
Unoxygenated blood comes through the Superior Vena Cava (SVC) and the
Inferior Vena Cava (IVC) to the Right Atrium (RA).
RA sends it to the Right Ventricle. The Right Ventricle then pumps it into the
Pulmonary Artery to go to both lungs to be oxygenated and return to the Left Atrium
(LA) via the Pulmonary Veins (SPV & IPV):
Two veins on the right side for the right lung and two veins on the left side from the left
lung.
The Left Atrium (LA) then sends the oxygenated blood to the Left Ventricle (LV) which
pumps the oxygenated blood into the Aorta to be distributed throughout the entire
body, from the brain to the feet.

Figure 1: Pulmonary Circulation

FATAL CANCER: JEREMY'S STORY

Jeremy was a fifty-year-old African American male smoker who was working as an electrician at Com Ed in Chicago. His main complaint was coughing, sometimes with blood stained sputum. He was referred to me by an internist on the staff of Mary Thompson Hospital, because an X-ray showed a central tumor located in the right lung in close proximity to the heart. When I went to see him in consultation, I found him sitting in a chair next to his bed, reading the Bible. I introduced myself and told him why I came to see him. He barely looked at me, as if he already knew me. I informed him that he would need a bronchoscopy with biopsy to determine whether the tumor was benign or malignant. He seemed unconcerned though, and responded by saying, "When do you want to do it?"

"Tomorrow," I said, going with his nonchalant flow.

He agreed, and continued reading his Bible. He was an easy going patient, and the bronchoscopy went smoothly. I could not take a biopsy, because the tumor was surrounding the main bronchus to the right lung. But I performed brushings and washings from the lining at the site of the tumor. Three days later the cytology report indicated that the cells from the brushings and washings were malignant, but did not specify the type of malignancy.

I went back to his room and tried to give him the bad news as tactfully as possible. I told him that the tumor was malignant but could be removed by surgery, though it might result in removing his entire right lung. He said,

"Doctor, do what you need to do," and continued reading the Bible. He barely looked up, and I was amazed at his calm, his poise and his faith in the face of such bad news. It seemed as if he believed that his prayers could overcome anything. And it seemed that he was surrendering his fate totally to God.

I continued by telling him that the surgery could be very risky, because of the closeness of the tumor to the heart. I was impressed by his response, "God will be with us."

His faith and courage touched me deeply. I knew that this operation was not going to be easy, because his tumor was central, involving the blood vessels returning the oxygenated blood to the heart (the pulmonary veins). It could be extremely dangerous, but I was determined to do my best for him.

His surgery started out as routine. We entered the chest by cutting into the space between the fourth and fifth ribs, and widened this space mechanically by retractors. Then we inspected the lung to find the exact location of the tumor and its consistency.

As expected, we found that the tumor was quite hard, and it extended to the pulmonary veins (see the diagram and description). To remove such a tumor would be difficult, hazardous or impossible. The pulmonary veins could be injured when dissecting them from the tumor, which would result in an uncontrollable and perhaps fatal hemorrhage. I was vexed by two thoughts: should I close the chest and tell him the tumor was inoperable, or attempt to remove it and suffer the possible consequences of serious complications?

This dilemma dominated my thinking, and for a moment it was tough to choose the best path. What led me to the second choice was the patient's confidence and faith. His words, "God will be with us," churned in my mind. The assured and serene image of my mentor, Dr. Cooley, was like a presence in the room. The next thing I knew, I found myself confidently stapling the bronchus and cutting it, and then doing the same to the right pulmonary artery. In that moment, we crossed a fail-safe point; the lung had to be removed. Now came the moment of biggest risk. Removing his lung with the cancer could not be done in the usual way of ligating and cutting the pulmonary veins. The only way to remove his lung with the tumor intact was to cut part of the healthy heart to which the pulmonary veins were attached, (Left atrium). It was a hard decision to make. Jeremy's life depended on this decision. I could either let him die from his cancer, or give him a chance to live by removing his lung with the tumor and part of the upper chamber of his heart. It was a choice between putting both my reputation and his life at risk, or simply closing him up and letting him die an agonizing death. I went for the riskiest option.

A radical pneumonectomy was Jeremy's only chance for survival. While this debate rattled in my mind, the words of one of my previous superiors suddenly hit me:

"In certain situations, to be an honest surgeon, you must be ready to risk your reputation in order to save a patient's life."

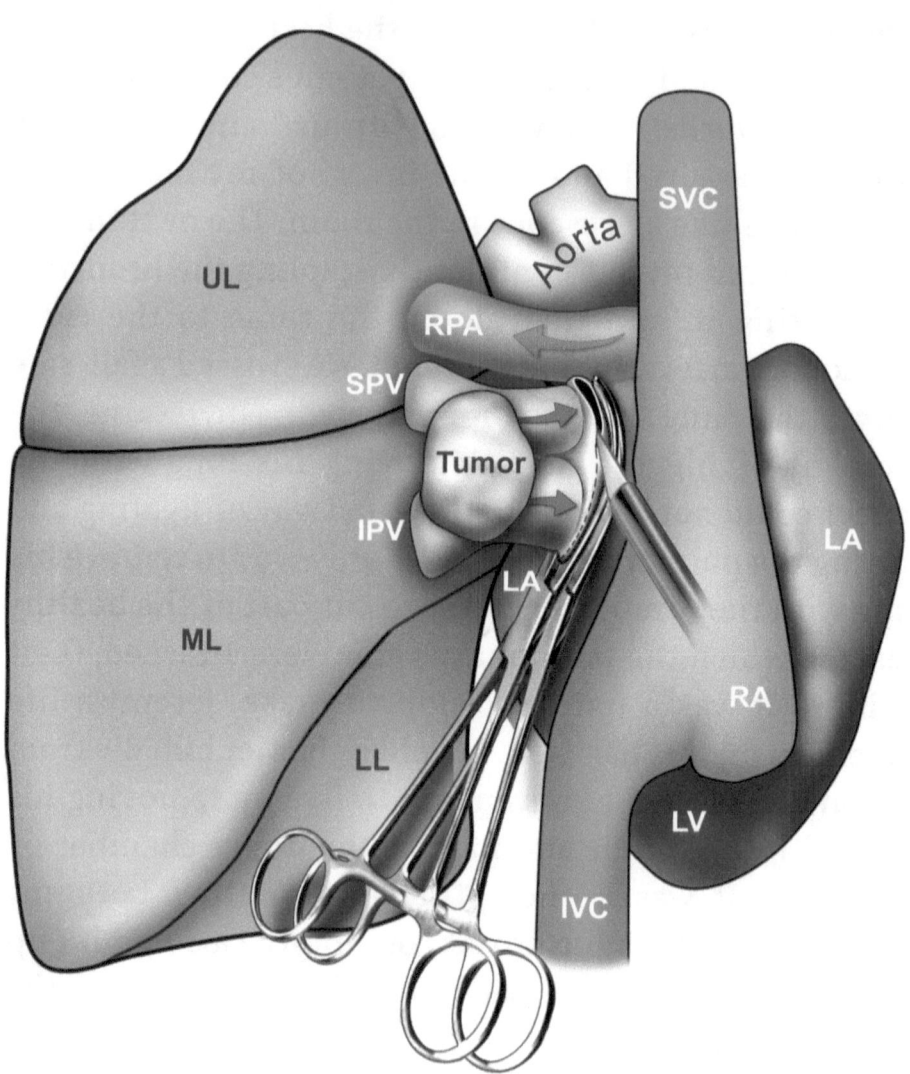

LOBES OF THE LUNG:
UL= Upper Lobe
ML= Middle Lobe
LL = Lower Lobe

SVC = Superior Vena Cava
IVC = Inferior Vena Cava

Figure 2

14

That ended my internal conflict. A spirit of courage, faith, and confidence enveloped me, and I found myself opening the pericardium and proceeding to remove his lung, along with its deadly cancer. After stapling and cutting the right bronchus and right pulmonary artery, there was no alternative but to remove the lung with the cancer and part of the left atrium where the involved pulmonary veins drained. With unshakable determination, I requested a Cooley atraumatic clamp and put it on the left atrium. I was going to start removing the lung with the piece of heart when my assistant, who was a fully trained vascular surgeon, asked to be excused. I looked at him and found him clammy and sweating.

The cancer covered the pulmonary veins which bring the oxygenated blood to the heart. To remove the cancer completely, I had to cut through the left atrium.

"What's the matter?" I asked him.

"If this clamp you put on the heart slips, the patient will exsanguinate to death. No one will praise us for taking such a risk. We'll be accused of being reckless and suffer a fatal blow to our professional reputations."

With the serenity, confidence and sense of humor which I had acquired from Dr. Cooley, I told him, "Cooley's clamps never slip."

I must add here that the anesthesiologist was actually supportive. He ordered more blood and told me that he was ready for any outcome. His positive attitude changed the atmosphere of the operating room to one of total confidence and to quell my assistant's anxiety, I added another Cooley clamp under the first one, and told him that if the first clamp slipped, which was a rarity, the second clamp would hold. This reassured him and he calmed down. I cut the heart containing the pulmonary veins distal to the two clamps and removed the lung with its cancer. It was then handed to the circulating nurse to take to pathology. I then closed the incision in the heart in two layers of sutures.

At this point, I started to open the first clamp bit by bit, and we were all happy to see that there was no bleeding.

My thoughts raced. When I start to remove the second clamp slowly, will the stitches hold? What will I do if bleeding occurs in the suture line?

I decided to reinforce it, but there was no bleeding. After removing the second clamp there was a cheer of triumph in the operating room along with a feeling that the operation was successful. I asked the pathologist if there was any tumor left in the cutting line by frozen section, and she reassured me that the tumor was completely removed with good safety margins. We all took a deep breath and proceeded to close the chest in the usual manner.

Jeremy's post-operative course was smooth and uneventful but the miracle did not end there.

The final pathology report stated that the carcinoma was of the *small cell type*, which is the most malignant kind of lung cancer. It's the worst, because its cells divide more rapidly before they mature. That's why they remain small and create a rapidly growing cancer. Many surgeons believe that by the time a small cell cancer is discovered, it has already spread in the blood stream, making it inoperable.

This was very disappointing to me personally, because I felt the risk was taken in vain and the patient would not be rewarded by the success of the operation. I was preoccupied about how to give such disappointing news to the patient, but when I started to tell him that it was a small cell cancer, before I even finished the sentence, he surprised me by saying, "Thank God it was small cell. If it was big I would've been dead by now."

I wondered, "Should I tell him that the small cell is the most malignant, or keep it to myself and let him enjoy the joy of his misperception?" Then, I considered the effect of mind-over-body and decided not to ruin his positive attitude, hope and cheer. Besides, there was nothing more to do.

God works in mysterious ways. Jeremy was fifty years old at the time of surgery, and according to literature, he should not have lived more than two more years. I offered to put him on disability, but he refused and went back to work as an electrician at Con Ed as soon as his wound healed. It was a pleasure to see him monthly for his general checkups, though his probable fate worried me. To our great surprise and satisfaction, there was no spread of the cancer, and he remained healthy. His only complaint was that his sex life had decreased. This showed me though that he still had zest for life, despite all he'd been through. He beat great odds.

Every year, I got a letter from the cancer registry asking if Jeremy was still alive, and if so, they needed to review his pathology slides for repeated re-inspections. They wanted to eliminate their doubts, and our pathologist always complied.

This went on for 15 years. Then, one day the patient brought me a small present and said that this might be his last visit. I became worried he had lost his faith, when in actuality it was at its strongest. He was now sixty-five and was going to retire to his hometown in Mississippi to live the rest of his life with his family in the atmosphere of his childhood. I always received a holiday card from him. Although he kept in touch, I missed seeing him because every time he came in for his checkup, the satisfaction of seeing him well, working, and happy, was a source of inspiration for me. It's the best reward a surgeon can have.

NOTE

Let us try to explain the factors that led to this patient's miraculous recovery, one which, by all expectations, should not have happened. I think his faith saved him and inspired me to do a hazardous radical operation to remove all the lung cancer which was stuck to the heart. This may have had a role in the cure. Maybe his belief that small cell carcinoma is better than large cell carcinoma had a most positive effect. Or perhaps, his salvation from that fatal type of cancer could be attributed to his constant prayer to an omnipotent power, higher than himself, to which he surrendered his fate.

PERSONAL REMARKS

In medical literature, there are reports of unexpected mortality after a pneumonectomy in patients who had no intra-operative complications. Subsequent experience has shown it's usually due to post-operative over-hydration. The usual 80cc's per hour of IV 5% dextrose in half strength saline, which is adequate for most patients are too much for a post pneumonectomy patient. The remaining lung is very sensitive to fluids, and any excess thereof will lead to pulmonary edema, which is very difficult to cure. Once this happens, patients are intubated, suctioned, and given diuretics. What generally follows is the onset of a series of post-operative problems. These however, can all be avoided. In my experience, it's best to dehydrate them slightly within the first 24-48 hours. After a pneumonectomy, patients don't suffer post-operative ileus, so they can be fed clear liquids orally by the second day after surgery. I generally don't allow them more than 50cc's IV per hour on the first or second post-op days. And on the third day, the IV can be discontinued, along with a progressive increase in their diet.

Though commonly done, I don't insert a chest tube after pneumonectomy, because the suction increases the already existing shift of the mediastinum towards the empty side. I'd much rather let the empty hemi thorax fill with fluid to balance the pressure coming from the other side. Moreover, without a chest tube, patients can ambulate easier and earlier. To diminish post-op pain, I usually cut the nerve of the space with which I entered the chest, or infiltrate it with a long-acting anesthetic.

RARE INTRA BRONCHIAL TUMOR

INTRODUCTION

We will discuss here, on the following pages, other tumors of the lung:

• Intrabronchial tumors

• Carcinoid tumors of the lung, which can be a sort of hemmorage, as in Gloria's story on page 28.

Granular cell myoblastoma of the bronchus (GCM) is a very rare tumor. Only forty-six cases have been described in medical literature. I had the opportunity to observe a patient suffering from GCM for approximately one year before she consented to surgery.

MARTHA'S STORY

Martha was a thirty-two year old, obese black woman admitted to Mary Thompson Hospital due to a persistent productive cough and recurrent lung infections with fever, dyspnea, and chest pain. Chest x-ray films revealed right lower lobe atelectasis. A fiberoptic bronchoscopy was performed on October 7, 1976, and revealed a patent, but irregular, lumen of the right lower lobe bronchus, with hypertrophy of the mucosa in a polypoid cauliflower fashion all around. Biopsy and brushing failed to yield a diagnosis, and the patient refused surgery.

On April 24, 1977, she was readmitted due to progression of symptoms. This time, plain chest x-rays and tomograms revealed a mass 4.5 x 4 x 3 cm in the right lower lobe area, which had not been present in the previous films. Martha still refused surgery, thus endangering her life.

An intrabronchial tumor occludes the lumen and prevents inflation of part of the lung below it; that part becomes deflated, collapsed, and its blood supply is compromised. Eventually it atrophies and becomes dangerously susceptible to infection. The woman was treading on thin ice.

About three months later, Martha was readmitted with a severe lung infection. Dreading the danger of her dire condition, she finally consented to a reexamination. The bronchoscopy revealed that the lumen of the right lower lobe bronchus was almost totally occluded with a fleshy mass that encroached on the middle lobe orifice. The severity of infection and difficulty breathing convinced her of the need for surgery.

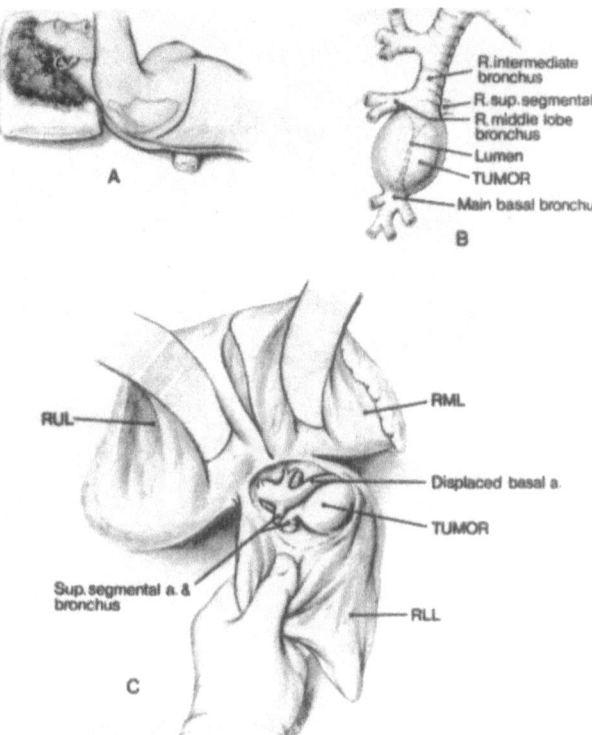

Figure 3: The operative appearance of the tumor during surgery showing the distortion of the hilar structures (part of the lung from which blood vessels enter the heart) by the tumor and displacement of the basilar artery. (Courtesy of Chest, 76:6, December, 1979.)

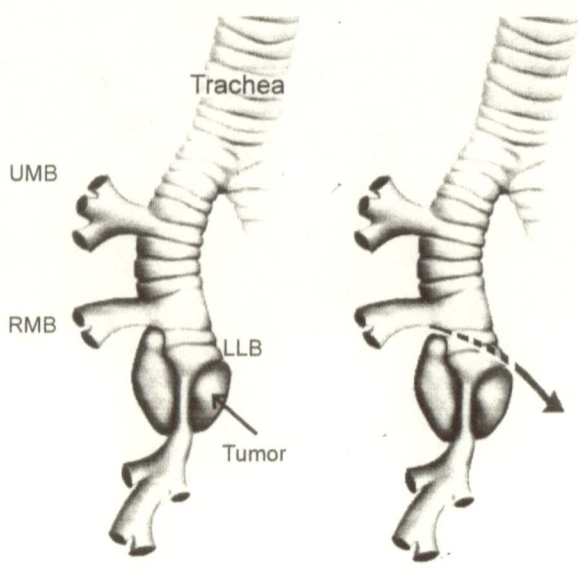

A.
Tumor occluding
lower lobe bronchus

B. Line of resection

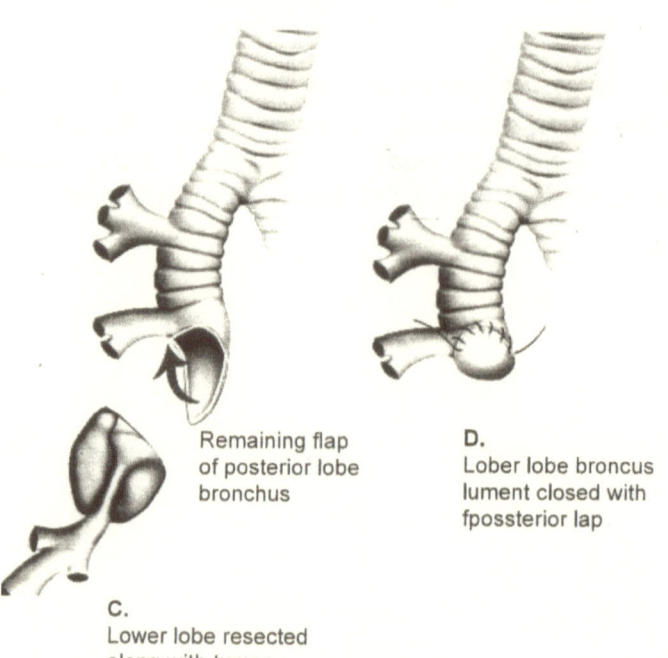

Remaining flap
of posterior lobe
bronchus

D.
Lober lobe broncus
lument closed with
fpossterior lap

C.
Lower lobe resected
alona with tumor

Figure 4

24

A right thoracotomy was performed, and a very hard mass was felt in the region of the right lower lobe bronchus. The size of the mass had grown tremendously since her first admission, which made surgery, and saving the middle lobe, much more difficult. However, a right lobectomy was safely performed and the middle lobe bronchus was covered by a flap to save it. With the infection removed, Martha made a gradual recovery.

DISCUSSION

Upon examining the tumor pathologically, it was found to be a very rare type of benign, but rapidly growing tumor called granular cell myoblastoma. Saving the middle lobe was of great importance to save the patient from respiratory insufficiency since she was obese, emphysematous, and needed every piece of healthy lung tissue.

COMMENTARY

Regarding intra bronchial tumors, whether benign or malignant, they should be treated surgically as soon as discovered, because they are a source of great complications, such as infection, bleeding, breathing difficulties and respiratory insufficiency.

CARCINOID TUMORS OF THE LUNG

INTRODUCTION

Some facts about these peculiar tumors known as carcinoids:

- They constitute 0.5-2.5% of all pulmonary tumors. They are relatively rare.

- They are more common in females than in males and in whites rather than blacks.

- They are usually centrally placed.

- The mean patient age is 40-50 years.

- They have no link to smoking.

- 80% of carcinoid tumors appear in a lobar bronchus (not in our case).

- They cause reduction in the size of the lung.

Hemoptysis occurs in more than 50% of these tumors, because they're highly vascular, though massive bleeding, as in this case, is rare. Less than 3% of all patients with a carcinoid tumor exhibit the signs and symptoms of what we call carcinoid syndrome. This is due to the tumor's secretion of 5-hydroxy-tryptamine (5HT).

Carcinoids are normally diagnosed by a CT scan. A routine chest x-ray shows vague symptoms of a lack of vascularity, and a decrease in the volume of the affected lung. The tumor could be hidden behind the shadow of the heart and that makes it difficult to detect with routine x-rays.

BLEEDING TO DEATH AND SUFFOCATING: GLORIA'S STORY

After making my usual rounds at Northwest Hospital, and preparing to leave, I was suddenly paged to the emergency room to see about a lady named Gloria, She was in her mid- 40's and had been brought in by her husband for massive bleeding from her nose and mouth.

Gloria was not a smoker, but I still wondered if this necessarily ruled out cancer. The emergency room doctor typed and cross-matched her, and she received saline, pending the arrival of cross-matched blood. Her hemoglobin was 9, and the husband said that it was usually 12 to 13. The woman's blood pressure was 80/60. She was gasping for breath and about to go into shock.

The first dilemma to solve was to find the bleeding source: the lungs or the GI Tract? There was no history of indigestion, though from time to time she'd cough up blood streaked sputum. This narrowed the problem to the lungs, but I couldn't tell if the bleeding came from the trachea or the actual lungs. And, if so, which lung? The chest x-ray revealed the trachea was opaque (filled with blood), but there were no filling defects, indicating a tracheal lesion. It also revealed some diffuse atelectatic areas of the right lung. The right lung was hypovolemic; the left lung was hyper-inflated. This suggested that the bleeding was probably coming from the left lung, but if so, from which part? The patient's condition was deteriorating.

There was no time for a CT scan or MRI. The patient needed emergency surgery to save her life. She was taken to the operating room, intubated and anesthetized. I tried to do a bronchoscopy, but all that I could see was blood filling the trachea and the bronchial bifurcation (the carina). The flow of blood was more than the scope could suction.

The only way to save this patient's life was to remove the part of the lung from which the bleeding was coming, but what if I removed the wrong side? The reasons for which I concluded the bleeding was from the right lung were vague to both me and the radiologist. While both of us suspected the bleeding was from the right lung, we could not conclusively confirm it. Such are the dilemmas that occasionally face a surgeon. At times like this, he has to totally depend on his *intuition*. Not knowing what lobe or segment the bleeding was coming from, I had to follow our hypothesis and do a right pneumonectomy in order to stop the massive bleeding, no matter which part of that lung it came from.

In the absence of a visible tumor on the x-ray, I was suspicious of a central carcinoid tumor. This is the only tumor of the lung commonly known to bleed, but there were no systemic signs of carcinoid tumors, such as flushing, fever, nausea, diarrhea or hypertension. Only 3% of carcinoids exhibit such generalized symptoms, but most carcinoids are extremely vascular and could cause coughing up blood (hemoptysis), but not to that degree. There was no time to hesitate. I told the anesthesiologist to give her left sided ventilation.

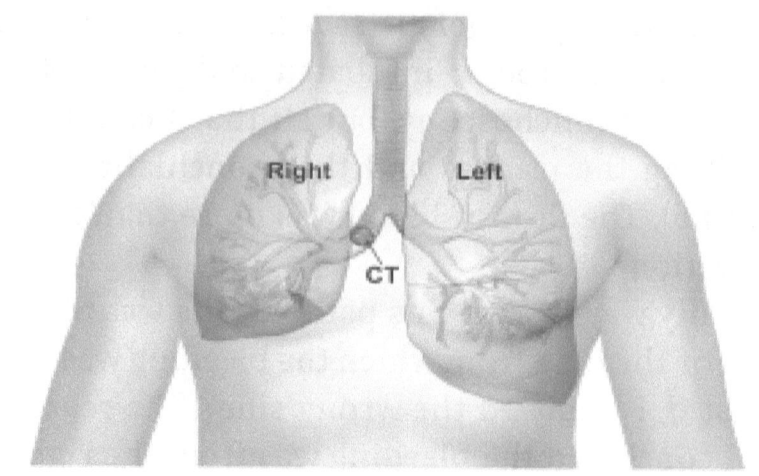

CT= Massively bleeding carcinoid tumor in the right main bronchi

Figure 5

A right exploratory thoracotomy was performed between the 4th and 5th ribs. I collapsed the right lung which exposed the hilar vessels. I then surrounded the pulmonary artery and two pulmonary veins with tape, but hesitated to ligate and cut them as I was not sure whether the tumor was in the right main bronchus or in a lobular bronchus. I dissected the right main stem bronchus with the idea of doing a bronchoscopy through it to locate the source of bleeding. Palpation of the lobes did not reveal any masses.

I opened the right main stem bronchus for about an inch, transversely using a scalpel, and before I could do any bronchoscopy I was able to see a fleshy mass, hard on palpation, and bleeding profusely.

Now I faced a new dilemma; it would take just a few minutes to perform a pneumonectomy and stop the bleeding, no matter which part of the lung it came from. However, that procedure could cause the patient to lose a lung.

The other option was a sleeve resection, (removing the part of the bronchus containing the tumor, and reattaching the lung). This would take much longer however, because I didn't know the extent of the tumor into the right main stem bronchus. Also, this might not be feasible at all if the tumor occupied all the right main bronchus.

The third option was to open the bronchus, remove the tumor, and close the bronchus. But leaving the base of the tumor might cause a recurrence. I also thought the tumor could be malignant; I had never seen a malignant tumor bleed so profusely. My intuition was that the tumor was a carcinoid, and a possibility of its recurrence was remote. I drew the tumor towards the bronchotomy, using an Allis Clamp, and used the bovie to resect it at its stump.

With the removal of the tumor the bleeding stopped completely, allowing me to suction the blood from the right lung. Then I did a bronchoscopy through the bronchotomy, and found no bleeding from any of the lobular branches. I closed the bronchotomy with interrupted sutures and covered the suture line with a pleural graft. I asked the anesthetist to withdraw the endotracheal tube and inflate both lungs.

A triumphant feeling washed over me, because the right decision had been made. The blood pressure stared to rise and the patient started to put out urine. I closed the chest in the usual manner, and thanked the anesthetist for being calm and for balancing the fluid intake and output.

In the post-operative period, we kept the positive pressure to a minimum for fear of a leak in the bronchotomy. The patient was very cooperative and we were able to wean her off the respirator within 24 hours. The next day, we got her out of bed to sit in a chair and started feeding her clear liquids. On the third day the chest tube was removed and we started to ambulate her. The patient, her husband, the referring physician and I were all overjoyed when we discharged her on the tenth post-operative day without the fatal hemorrhage that almost suffocated her.

A competent, cooperative, calm anesthesiologist is of great value to the surgeon in such a catastrophic situation, and that contributed to the success of the surgery.

BREAST CANCER

"Don't accept any so called surgical principles as facts. They are often based on faulty impressions. Keep asking why? If? How? I have assumed a rather disrespectful attitude toward many concepts and have been willing to change them. This has often paid off!"
—Denton Cooley, M.D.

INTRODUCTION

Breast cancer is the most common cancer in women, and there have been important progressive changes in its operative and post-operative care, from radical Halsted mastectomy to simple lumpectomy. Other care options include: cutting down on unnecessary radical axillary lymph node dissection and vigorous modalities, radiation, chemo, and confining their use to metastatic cancer only.

Such information is necessary for women to make an informed choice to refuse any treatment imposed on them, especially if they doubt its benefit or necessity. The patients who were cured from breast cancer wanted their experiences published to help others in making their own decisions when searching for a cure.

In the tradition of not blindly following what others do, I personally think that if there are no palpable lymph nodes in the axilla, then a complete lumpectomy is all that is needed. The patient can consider herself cured, and forget she ever had cancer. After that, she can just go for routine checkups.

If there are palpable lymph nodes in the axilla, I prefer to remove them, and not do a radical lymph node dissection, which results in an irreversible, painful lymph edema of the arm. I've also found that non-palpable lymph nodes rarely carry any malignant cells. For those rare unpalpable lymph nodes which may carry metastatic cancer, I rely on irradiation of the axilla to destroy them.

My patients are too valuable to me after I rid them of cancer by a simple lumpectomy to then have them suffer "lymph edema," which is not only unsightly and painful but incurable. To expose women to such unnecessary agony is unacceptable to me. Furthermore, to irradiate a breast after a complete lumpectomy with a safety margin is overkill and can be harmful.

Unnecessary irradiation of the breast can predispose the patient to other cancers and lower the resistance of the patient to recurrence and distant metastasis. The same goes for not irradiating the axilla if there are no palpable lymph nodes. Should any palpable lymph nodes appear at a later date, they can be removed and the axilla irradiated.

Radiation and chemo are potent therapies that carry many side effects; these treatments should be reserved for recurrences or metastasis. If wasted on "prophylaxis," they won't be effective when needed. A protocol which tells a woman: Lumpectomy + radiation + chemo = mastectomy is returning to the old tendency of overtreatment. It's deceptive, harmful and, in my opinion, cannot be justified. What I would recommend after a radical lumpectomy is to ask the pathologist to check for safety margins and estrogen receptors. If positive, I recommend Tamoxifen; one tablet daily, for five years.

In the following pages, we will discuss lumpectomy instead of mastectomy, and the case against following the usual protocol. This will be demonstrated by the three stories that follow.

REFUSED THE PROTOCOL: HELEN'S STORY

Helen was seventy years old when she and her husband came to see me with a mammography report and a positive biopsy report for a cancerous tumor in her left breast. Doctors had recommended a mastectomy, but the patient refused, and came to me for a second opinion.

On examination, I found that she had a hard mass easily felt by the palm of the hand. The size of the tumor was about 5 cm in diameter. There were no palpable axillary lymph nodes. When I suggested that lumpectomy would, in my opinion, be sufficient to treat her tumor, she accepted it without hesitation. And this lady was my kind of patient, one who had the courage to deviate from the "routine" if she didn't go for it. This is the mood that inspires success.

I performed a complete lumpectomy on her and she was discharged the next morning. Her post-operative course was smooth and her wound healed well. On the seventh day, the stitches were removed and I told her about the commonly accepted "protocol" of radiation and chemotherapy. I won't ever forget her response to that.

"I'm seventy years old, and you took the tumor out. That's all that I needed. I don't want those other agonizing therapies. I'm happy as I am and if I die, I die."

She loved music and enjoyed it as relaxation therapy. She and her husband had a monthly subscription to the Symphony in Orchestra Hall in Chicago and asked me if I could accompany them whenever possible. I too appreciate great music, and it was a pleasure to accept.

Helen lived until the age of 93 and died peacefully in her sleep, never suffering any recurrences.

Helping my patients in different ways, beyond the call of duty, was not unusual in my practice. It always resulted in good doctor-patient relationships, which I feel, are of the utmost importance.

NO MASTECTOMY, NO PROTOCOL: NANCY'S STORY

Nancy is a fifty year old dentist and mother of three boys, who came to me in 1985, bringing with her a biopsy report of breast cancer taken at the Mayo Clinic. Her older sister had suffered from breast cancer and underwent a mastectomy and prophylactic mastectomy on the unaffected side. This was followed by the protocol, (irradiation and chemotherapy), and she died within five years.

Upon examination, she was overweight with large pendulous breast, and had a hard mass in the upper quadrant of the left breast about the size of a grapefruit. A metastatic workup revealed no further cancer spread and no palpable glands in the axilla.

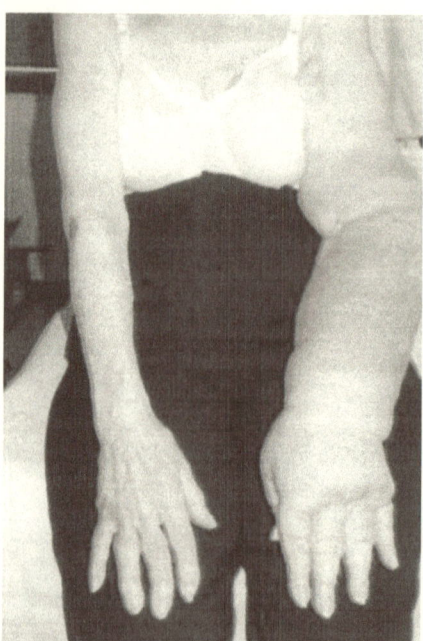

Figure 6: Swollen painful arm following radical axillary lymph node dissection.

Since she refused a mastectomy and because of the big size of the tumor, my wife Pat, a Registered Nurse, and I took her to the radiology department and had the radiologist mark the boundaries of the tumor with wire under mammography x-ray guidance. The next day, surgery was performed and the tumor was excised with a centimeter or more beyond the wire. It was sent to the pathologist to examine for safety margins and estrogen receptors. The tumor margins were free of cancer and the tumor itself was negative for estrogen receptors. The rest of the breast tissue was sutured together and the skin closed in the usual manner.

Because of the size of the mass, the left breast appeared slightly smaller than the right but amazingly, it filled up with time and a couple of years later, they were almost equalized in size. Every time she went to the Mayo clinic for routine checkups, they asked the same question, "No radiation, and no chemotherapy?" She replied, "Is there anything wrong with that?" The doctors graciously answered, "You cannot argue with success."

Fifteen years later, a mass the size of a lemon appeared in the left axilla. I suggested resecting the mass, only because there were no other palpable masses. But this time she went along with the Mayo Clinic recommendation, whereby they performed a radical lymph node dissection on her left arm. To everybody's surprise, the mass turned out to be a lymphoma and not breast cancer metastasis. This led to further investigations for lymphomas elsewhere in her body. A CT scan of the abdomen revealed multiple masses in her abdomen. At that time, she refused my personal treatment for lymphomas and opted for chemo and irradiation.

Her arm is now swollen and gives her constant pain. Ironically, if they knew it was a lymphoma and didn't follow the conclusion that it was breast cancer metastasis, they would have excised or irradiated the mass alone and saved the patient the agony of radical dissection.

To date, twenty years later, there's no evidence of recurrence of the breast cancer. But the treatment of the lymphomas with irradiation and chemo does not seem to be very successful. There are times she suffers from agonizing pain, and at one point she suffered temporary paralysis of a leg because of the pressure of the lymphomas on the spinal nerves.

Patients treated with my technique lived a longer and painless life. It's unfortunate that the statement, "lymphomas are not a surgically treated disease" is being repeated throughout medical literature. It goes unquestioned, and the routine post lumpectomy protocol for breast cancer goes unchallenged.

In spite of her suffering, and the remissions and intermissions of her lymphomas, Nancy continues to run her dental clinic. She appreciates and laughs at every joke, never misses a party, and goes to church almost every Sunday. Her faith and attitude keep her going. She is a shining example to be followed and deserving of our respect.

SHE OPTED FOR THE PROTOCOL: CHERYL'S STORY

Cheryl L. was my wife's cousin; thirty-three years old and a UPS truck driver. She had a history of a few car accidents, sometimes with open bone fractures. Cheryl complained that when elevating her arm, as when combing her hair, reading a paper or driving a car, her arm would go numb.

She was an attractive young lady with stable vital signs and a slight limp. I had her perform an elevated arm stress test, which involves elevating the arms at right angles and opening and closing the fists. Within a minute of beginning she felt pain and numbness in her arms, and her palms became pale. These are typical symptoms of *thoracic outlet syndrome* (TOS), and she had to undergo a complete first rib resection.

Cheryl came in for removal of stitches on the seventh day, at which point the symptoms of numbness and pain in her left arm completely disappeared. After removing the stitches I could feel a small swelling at the end of the incision, close to the axillary tail of the left breast. It looked like a fibroma or a small stitch abscess.

By mere act of intuition, I excised the small pea sized nodule and sent it in for pathologic examination. I really doubted that it was necessary, except for an inner voice that told me to do it. Sometimes those intuitive urges to do something cannot be explained, yet prove invaluable.

To my great astonishment, the report indicated a small ductal carcinoma of the breast with safe margins of excision. To be very cautious, I readmitted Cheryl to the hospital and made a wider excision around the previous site and made sure no cancer cells were left behind.

Since there were no palpable axillary nodes, I considered this as curative. The new wound healed completely, and two weeks later she went back to her job as a bank teller. It was lucky the tumor was discovered early and could be totally removed before it enlarged or spread. She was very content.

DISBELIEF AND MISTRUST

I was astonished when four months later Cheryl appeared in my office and insisted on having a copy of her records. Surprised by her insistence and change in mood, I asked her what happened. She told me that while watching television she came across an ad about "breast cancer specialists." She had gone to see them and they told her that, "a wide excision is not enough, and that she should have had a mastectomy."

My suggestion that since the tumor was very small when discovered and that further excision around it showed no more cancer, she, in my opinion, was cured. But this fell on deaf ears, and she went in for repeated courses of irradiation and chemo. A few years later, she developed a persistent pleural effusion and was sent home with a chest tube. Sadly, she remained bedridden until she died at the young age of forty-five.

NOTE

If a lump is discovered by patient palpation, mammography, or ultrasound of the breast, in my opinion, the first step is to find a surgeon with experience in breast cancer to perform a *needle biopsy.* If the mass is cystic, fluid will seep into the syringe and the mass will collapse. If the tumor is solid, my recommendation is to perform a wide *"excisional" biopsy.* If the tumor is proven to be benign, this takes away concern on the patient's part regarding malignant transformation. If the tumor turns to be malignant, then nothing more should be done.

Personally, I would not recommend an open biopsy, (taking a small piece of the tumor for examination), because the biopsy could be taken from a piece of the tumor which is not malignant and give the patient a false sense of security. If positive, it exposes the patient to a second operation.

As mentioned before, if there are no palpable lymph nodes in the axilla, that's all that needs to be done. If there are palpable lymph nodes, they should be excised and the axilla irradiated. I don't recommend radical lymph node dissection at this stage, because it can lead to unnecessary, painful, irreversible lymph edema (severe swelling) of the arm for the rest of the patient's life.

At the present time, I feel no mastectomy is justified. It's disfiguring and takes a lot of psychological adjustment to cope with by both the patient and partner. It's a thing of the past and is very rarely indicated. Post mastectomy, reconstructive procedures are never a good substitute for the breast itself. The rare exception is when the cancer is neglected and fungated through the skin.

NOTE

When I was a resident, the usual procedure for any cancer of the breast, regardless of its location, was the radical mastectomy (*Halsted's operation*), which consists of removing the breast, the muscles under it and all the lymph nodes in the axilla. To me, this was a cruel operation which I detested. To see a woman with palpable ribs and no breasts appeared very disfiguring to me, and I highly doubted its benefits. For these reasons, I scanned medical literature and discovered that simpler operations were performed in other countries, especially in England, with better results in regards to longevity and recurrence of the disease. So, I arranged a symposium to review the subject of breast cancer.

I reported results published in British medical literature which concluded that simple mastectomy was better than radical, in all aspects. I further used the fact that if the cancer was in the medial quadrant, the metastasis would be in the internal mammary lymph nodes which are inside the chest. How could this justify an axillary node dissection or even mastectomy itself? I argued that if we were going to use irradiation for the deep nodes following mastectomy, why wouldn't a lumpectomy alone be enough for tumors in this location? Furthermore, I suggested that, if the tumor is in the outer quadrant, there is no lymph node metastasis, and simple mastectomy is enough though a simple lumpectomy would be just as successful.

You cannot imagine the backlash I encountered. The fact that lumpectomy could have the same benefits as a mastectomy, without disfiguring a woman's body, lowering her self-esteem, and requiring a lot of psychological adjustment for both her and her partner, did not move them. It's amazing how hard it is to suggest something new. At that time the usual and customary method to treat breast cancer of any kind was radical mastectomy (the Halsted's operation). Its success was judged by survivors of five years. My superiors lived by that standard as portrayed by statistics in the literature. They thought my aim for supporting lumpectomy was purely aesthetic and psychological, but as far as longevity, it might not live up to the accepted five year survival goal. They opposed my suggestion and said: "You are a young emotional surgeon, and when it comes to cancer, life is more important than beauty."

Ten years later, Dr. George Crile from the Cleveland Clinic proved me right when he published a paper with a double blind study that proved lumpectomy is superior to mastectomy with regards to both longevity and quality of life. It was a great compliment for me when my former boss, Dr. M. Deanzman sent me the article with a handwritten note: "Josh, you were ahead of your time."

In my forty-year career as a surgeon (fifty total in medical practice), I rarely performed a mastectomy, and then, only when the tumor was neglected and fungated outside the skin of the breast.

At the present time, typical protocol is to follow lumpectomy with radiation and chemo therapy. But even the value of that protocol is questionable, as evidenced in the stories of the patients who had it and those who refused it. In my opinion, the protocol should not be routinely followed until its value is proven by double blind studies.

"With radiation after lumpectomy my breast gradually got flatter and flatter until it looked as though there has been a complete mastectomy. My breast degraded gradually losing more and more volume until it became non-existent." (Suzanne Somers. **Hidden Side Effects of Radiation**. *Journal of Life Extension*, December 2011, page 39.)

MILESTONES IN CANCER TREATMENT

INTRODUCTION

I could never forget a traumatic experience I had when I was a resident. One of my relatives had vague abdominal pains, and didn't respond to the usual medication. Finally, she accepted surgery to do a laparotomy to find out the cause of her symptoms.

Upon opening her abdomen, I found it full of metastatic swellings. I called my boss for consultation and he recommended I close the abdomen and send her to oncology for chemo and irradiation. In one month she was dead.

Such a distressing experience repeated itself a few times. Since then, as it is my custom, I kept asking myself over and over again if anything could be done to change such a gloomy prognosis. This kept haunting me until an opportunity inspired me while operating on patient Rose.

METASTATIC CANCER CURE: ROSE'S STORY

Rose was a forty-two-year-old African American police guard in a suburban prison. She came with her mother for whom I had performed a lobectomy for lung cancer several years before. She brought a CT scan of the abdomen showing multiple abdominal masses. Her primary complaint was persistent: vague abdominal pains. She gave a history of a total hysterectomy and bilateral salpingo oophorectomy two years before, with vague indications.

She was admitted for an exploratory laparotomy with biopsy of one of the masses. This revealed twelve masses of varying sizes; from a lemon to a tangerine, all hard in consistency, with some stuck to the liver, spleen, bladder, mesentery, aorta, etc. There was no fluid in the abdomen.

At the thought of closing her up after taking a biopsy, and sending her to die, as when operating on similar cases in the past, a gloomy feeling came upon me. I spent a few moments asking God if there was any way to give her a better prognosis. Not knowing at the moment what the answer was, I proceeded to take a wedge biopsy from one of the masses. *As soon as I incised the capsule of one of those tumors, the core of the mass protruded. The idea came to me to "scoop" it from the capsule completely and send it to pathology. To my surprise, there was very little bleeding inside the capsule, and it was easily cauterized.*

While waiting for the pathology report, I proceeded to perform the same procedure on the rest of the masses. With the capsule stuck to vital organs, the danger of a total excision was avoided. With a sudden wave of faith and

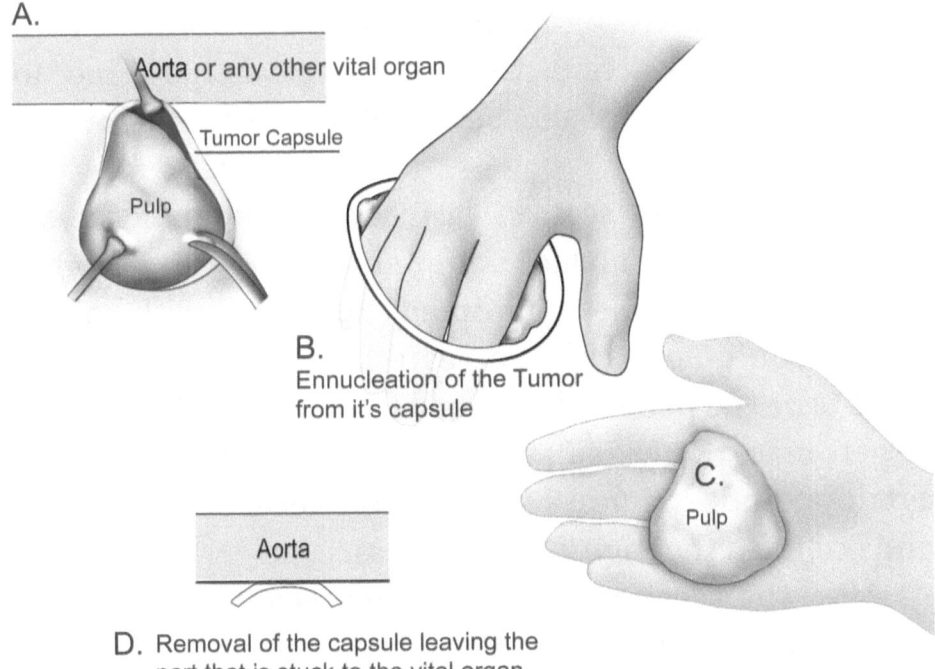

Figure 7: A new method to treat metastatic cancers and malignant lymphomas.

confidence, I proceeded to excise the edges of some capsules with bovie, leaving a small patch on the vital organ to which the mass was attached.

This did not please the medical director of the hospital who expected me to do just a little biopsy. It did, however, please the oncologist who said, "This is a great debulking of the tumors, making irradiation and chemo much more effective."

The pathologist and oncologist disputed the origin of these tumors, but both of them agreed they were malignant and mostly ovarian.

Upon closing the abdomen, it appeared much cleaner and nicer after removing the intrusive masses. I had a feeling of satisfaction that my hope helped me to persevere and that I did something of benefit for the patient.

After the first session of chemo, the patient was discharged feeling elated. I offered to apply for disability for her but she refused. She opted to continue with her job as a prison guard and two years later got married. For some time the patient developed fluid in the cavity of the largest mass. Repeated aspirations showed no malignant cytology, and four years later, I took her back to surgery and occluded the cavity of that mass.

After the second operation, there was no more fluid and the patient visited me every three to six months. Then she chose to be checked at a hospital closer to her home. Her mother, who used to come to me for routine medical problems, assured me that she was doing okay fifteen years after surgery.

Was this a serendipitous discovery? Was it a hunch or a response to my prayer for guidance?

Since then I repeated this procedure over and over again on this kind of cancer and lymphomas. The results are very satisfactory for longevity and a painless quality of life.

NOTE

I think this technique is a **milestone** in the treatment of metastatic cancers and malignant lymphomas.

LYMPHOMAS

INTRODUCTION

Lymphomas are neoplasms that arise from lymphoid tissues. They are divided clinically and histologically into two types: Hodgkin's and Non-Hodgkin's.

HODGKIN'S LYMPHOMAS

Rubbery enlarged lymph nodes that usually appear in the neck or supra clavicular fossa and on x-ray appear like large mediastinal masses which are surprisingly asymptomatic. Occasionally they are accompanied by a dry cough. If left alone, they usually spread from one node to the next and enlarge within a short period of time. Their specific diagnosis is possible through biopsy, which shows a very special type of cell called Reed-Sternberg. The Hodgkin type has a much better prognosis than non-Hodgkin and its treatment is much more straightforward—it also depends on the staging.

STAGING OF HODGKIN'S

Stage I: A single region is involved.

Stage II: Two regions are involved and are above the diaphragm.

Stage III: Two sites are involved and are below the diaphragm.

Stage IV: The liver or bone marrow is involved.

Ninety percent of patients with stage I are cured by radio therapy alone; stage II is cured by radio therapy followed by chemo therapy.

NON-HODGKIN'S LYMPHOMAS (NHL)

NHL is histologically characterized by a 70% proliferation of B-cells; the remaining 30% are composed of T-cells.

Incidence: The occurrence rises with age. At the present time, they amount to twelve new cases per 100,000 people per year with slight male dominance.

Etiology: Viral—They're late manifestations of HIV, EBV, and herpes.

Finally, some lymphomas are associated with specific chromosomes, e.g. follicular lymphomas.

HOW NHL DIFFERS FROM HODGKIN'S

They are more widely disseminated and can appear as an enlargement wherever any lymph nodes are present.

Symptoms are more drastic. The lymph node enlargement is often associated with weight loss, sweats, fever and itching. It usually progresses to hepatosplenomegaly with abdominal masses or bone marrow involvement, at which point it is considered leukemia.

MANAGEMENT:
LOW GRADE NHL VS. HIGH GRADE NHL

Low grade NHL is treated by radio therapy followed by chemo therapy. If the intensive chemo therapy produces toxicity that interferes with the quality of life, in my opinion, the lesion should be surgically excised so the dose of chemo therapy can be reduced.

High grade NHL calls for immediate treatment at presentation because the symptoms are more pronounced and the disease is more diffuse. Oncologists usually start with four IV regimes, (cyclophosphamide, doxorubicin, vincristine and prednisolone). If the tumors are bulky and result in compression syndromes, radio therapy is added.

In my experience, these modalities usually do not result in significant size reduction of the tumors. In spite of the current treatment, abdominal lymphomas can still multiply and press on the spinal nerves causing severe pain or paralysis of a limb. In this case, the oncologists increase the doses of chemo and irradiation resulting in severe suppression of the immune system, which increases the likelihood of death from pneumonia or related infections.

LYMPHOMAS SAFELY TREATED:
MR. CHAMBER'S STORY

The patient in this story had a high grade non-Hodgkin's lymphoma with many recurrences.

Mr. Chamber was a forty-five-year-old milk man when he was brought to me in 1974 by his wife because of a painless mass in his right axilla. His vital signs were stable and the mass itself was the size of a tangerine, hard and irregular, growing significantly in size within six months. The blood count, chemistry, and bone marrow biopsy were completely normal. The sedimentation rate (SED Rate) was a little bit raised; LDH was normal; liver and renal function tests were normal. His chest x-ray and CT scan of the abdomen were negative for masses and the mass was excised. The pathological exam revealed it was a malignant lymphoma with a predominance of B-cells.

After excision, he received chemo and irradiation for six weeks. The patient returned to his job and for the next three years was free of symptoms or recurrences.

In 1977, he returned with a fast growing tumor in his left groin. It was treated in the same manner, but over the next five years he had recurrences in his right axilla and right groin which were treated similarly, but never in the same place. In 1982, he came with symptoms of nausea, indigestion and constipation. Abdominal examination revealed *splenomegaly*. I ordered a CT scan of the abdomen, which showed two masses in the upper abdomen and three in the lower.

A laparotomy revealed a spleen three times the normal size; lymph nodes the size of tangerines attached to his liver and stomach; three masses attached to his mesentery, bladder and rectum the size of oranges. I knew that no amount of chemotherapy or irradiation could dissolve these masses.

Deviating from what is taught in medical books, I opened their capsules and enucleated the tumor tissue from inside, and then excised the edges of the capsules, leaving only a small part stuck to the organ to which it was adherent. Then I proceeded with a splenectomy.

The surgery was without complications and the patient had a smooth recovery. Three weeks later, I sent him back to the oncologist whereby he received eight doses of chemotherapy one month apart.

At this stage, I got him to retire from his work and spend the rest of his life with his family peacefully, which he did for the next six years without further complications.

In 1988, fourteen years later, he developed fever with bronchitis. His wife took him to a suburban hospital where he received IV antibiotics. Because of his history, the admitting physician called for an oncology consultation and without contacting me, started him on a high dose of chemo, which, in my opinion, was not needed, as there were no masses in the abdomen by palpation or CT scans. This lowered his resistance and in a few months he contracted pneumonia from which he died a painless death.

FOR COMPARISON

Around this time, my brother-in-law developed a swelling in the right side of his neck. It turned out to be a malignant lymphoma. CAT scans of the abdomen revealed an aortic aneurysm and scattered masses which proved to be lymphomas. On exploration of his abdomen the surgeon excised the spleen, the aortic abdominal aneurysm and a biopsy of one of the masses. When we found that the biopsy revealed a lymphoma, following the dictum that *"lymphomas are not a surgical disease"* they closed the abdomen and ignored them. Post-operatively he was given eight monthly sessions of chemo, but during this treatment he complained of severe abdominal pains which led to him being admitted several times to the hospital for sedation and analgesics. In the ninth month, the pains were so sever my sister called me at two in the morning to give him a Fentanyl injection. A year after surgery he started vomiting and developed diarrhea. He was admitted to the hospital for dehydration and pain. Within a few days he died an agonizing death. That emphasized my conviction that chemo alone cannot dissolve or cure lymphomas. They are too big for the chemo to dissolve and I am glad that I was guided to a way that could decrease the suffering of such patients and prolong their lives.

It is worth repeating that lymphomas are not surgical targets. Maybe this is because of their attachment to vital organs which make their excision hazardous. Having seen such patients in agony inspired me to discover a way to remove these tumors without danger to major organs. Such debulking makes chemo more effective, as in the case of metastatic cancers.

When a specific treatment fails we must ask ourselves, *"Is there a better way?"* We must look *"outside the box,"* and think of new ways. If we think deeply and persistently, a new method will be inspired to us by chance (serendipity) or planning (discovery). That's how medicine progresses. We owe this to our patients and our profession.

OVARIAN CANCER WITH METASTASIS

INTRODUCTION

While most cancers occur at older ages, ovarian cancer is more common in young patients, usually below the age of thirty and very rare in post-menopausal women. The methods of detection are a pelvic examination, and if suspected, a C-125 blood test, followed by an ultrasound of the pelvis. When there is a history of ovarian cancer in the patient's immediate family, it is very important the siblings be vigorously tested, as the risk is higher. Ovarian cancer normally occurs in 1 out of every 50,000 women and if a mother and daughter have it, medical literature reports that the chance of the second daughter getting it is fifty percent.

A woman with a family history of ovarian cancer should be submitted to a cancer workup early in her life. Some gynecologists and their patients prefer to remove the ovaries immediately after delivering the desired number of children even if they don't show signs of cancer transformation.

On a more cheerful note, only 10% of tumors of the ovaries are cancerous. A patient should not be terrified if she is told that she has a tumor or cyst in the ovary since early diagnosis increased the chances of cure. Recently small ovarian cysts or tumors can be removed laparoscopically, and the patient can leave the hospital the same or following day. Patients on the pill or with early menopause are at lower risk of ovarian cancer since these factors decrease the number of ovulations. Other factors which decrease the risk are: a diet low in fat and rich in fiber, weight control, and avoiding the use of alcohol and tobacco.

Ovarian cancer is the fourth most common cause of cancer deaths in women. Unfortunately, it cannot be discovered early unless the patient has a family history of cancer that warrants the screening tests. The incidence of ovarian cancer in women without family history is only 15 per 100,000 women. When discovered late, only 15-20% survive for five years. If discovered early, 40% may survive up to five years.

Enlarged ovaries may be discovered upon pelvic examination, but the majority of them are not cancer. If a woman has ovarian cancer in her family, she should be subjected to ultrasound or CT scan of the pelvis at least once a year.

OVARIAN CANCER—A SURVIVOR'S CONTRIBUTION: VIRGIN JULIE'S STORY

Julie was a beautiful nineteen-year-old Caucasian woman who came to see me in 1989 with persistent abdominal pain on the right side. Her WBC (white blood count) was high and a diagnosis of acute appendicitis or periappendicular abscess was entertained.

Julie was taken to surgery, whereby the appendix seemed normal, but the right ovary was enlarged and hard in consistency with an ugly irregular surface. I sent a biopsy for frozen section, which, to my surprise and chagrin, revealed malignancy. Although it was late at night, the oncologist and pathologist of the hospital were called to look over the slides. It was three in the morning when they advised me to close the abdomen and wait for the final report.

Three days later when the final report came, I had a conference with the pathologist and oncologist. They decided I should re-operate on her, do a total hysterectomy, remove both ovaries, and excise the omentum (a pad of fatty tissue covering the intestine). I had to be careful not to excise big masses and expose the young girl to a serious hemorrhage. When we went to take the patient's consent, it was heartbreaking to tell her what we had to do. The poor girl burst into tears and said, "That means I'll never have children? I'll be menopausal while I'm so young and still a virgin?"

It was with a heavy heart that I took her to surgery two days later. I performed a total hysterectomy, removing both ovaries and the greater omentum which contained numerous enlarged lymph nodes. I removed her para-aortic lymph nodes. There were three masses ranging in size from an egg to a tangerine. One was *intra-mesenteric*; the other was attached to the *vena cava*, and the third to the *ureter*. Without exposing her to any danger, I opened the capsules of these metastases and scooped their pulps leaving part of the capsules attached to the organs in danger. I was determined to clean her abdomen from any cancer. Surgery took four hours without any complications.

At the same time, my wife, an R.N. in OB and neonatal care, was hospitalized following a surgical procedure. When Julie left the operating room, by coincidence, she was placed in the same room as my wife who counseled her and encouraged her to have a positive outlook toward the future.

When the wound healed, she was put on chemo therapy for a week every month for six sessions. During these months, I saw her regularly and told the girls in my office her story and they referred to her as "Virgin Julie." She heard this and was not offended.

I had great sympathy for Julie and would have loved to have her remain as my patient, but she happened to belong to an HMO that insisted to keep her future care under its own doctors so I lost her for follow up.

THE BIG SURPRISE

In 2005, sixteen years later, just before leaving my office, my receptionist came to tell me there was a lady waiting for me. When asked for her name and reason for her visit, she was told, "Just tell Dr. Salvador, I am Virgin Julie."

I couldn't believe my eyes when I saw her. I had no idea she was still alive. I hugged her and kissed her cheeks. Her features had changed some since 1989, but she was still beautiful. I told her I was happy to see her, but asked what brought her back after such a long period of not being under my case? She asked me for her records. I asked what she wanted them for, and I was so proud of her when she told me that she was a counselor for women who had ovarian cancer and that she was a living example to encourage them and give them hope. This was her way to give back for the blessing of God keeping her alive and cancer-free all this time. It cheered me when she told me that she was dating a nice man and contemplating marriage, with the idea of adopting a child.

Her departure from my office was a very emotional moment for me. But it was very rewarding to know she was still alive with plans for a bright future, as well as helping others. I'd have loved if she would've come to see me again after ordering her records from the hospital. She could have gotten them by herself, but I think she wanted to surprise me and show her gratitude. I wished her a long happy life, and after she left I was so emotional I had tears of happiness in my eyes.

My memory of her will never fade. There's no bigger reward for a surgeon than that.

NOTES

In the past, ovarian cancer was considered a death sentence. It was the most fatal gynecological cancer because the symptoms of subtle ovarian cancer could not be diagnosed until it had spread, making it more difficult to treat.

The birth control pill lowers the risk of ovarian cancer, because it reduces the number of ovulations. Some improvements happened when the surgeons started to remove the ovaries, fallopian tubes and uterus. They were to deal with metastasis to avoid the risk of cutting additional organs, and also to erase the idea that the risk outweighed the benefits. They relied on chemotherapy to deal with metastasis.

With the following procedure women could live 18-22 months: aggressive surgery by adding removal of the metastasis and advances in chemotherapy, made this disease manageable. My procedure of enucleating the cancer from its capsule made response to chemo more effective. Removing every last bit can make a difference in survival and cure rates. Added to this were improvements in medical therapy. For example, the drug Avastin could stop the blood vessel formation that feeds the cancer. Injections of *PARP* inhibitors in the abdominal cavity also made a difference.

Early diagnosis increases the chance of cure. Women with ovarian cancer history in the family should be screened every six months by ultrasound of the pelvis and/or abdomen, and added CT scans if the ultrasound was positive.

Blood test for BRCA 1 and BRCA 2 (genes) would reveal the genetic predisposition to cancers of the breast and ovaries, and could lead to preventive excision of the organs before they acquire cancer.

THE MIRACULOUS DISAPPEARANCE OF DISEASE

"The Divine Mind and its actions are beyond the human thought and its limitations."

—Charles Baker

WARTS: AN INTRODUCTION

Warts are ugly, painful papillomatous growths of the skin. They can occur anywhere, but are most common on the hands. They are typically the result of infection with the human papillomatous virus, (HPV), of which there are many subtypes. They're transmitted by direct contact with the virus and are extremely contagious.

They are treated by dermatologists with ointments containing salicylates and lactic acids, but such treatment needs to continue for several months before its effect becomes apparent. Some are treated with cryotherapy, using liquid nitrogen. This treatment is accompanied with significant pain and needs to be repeated at intervals of 2 to 4 weeks and can lead to significant complications, including tendon rupture. Other modalities include systemic retinoid or sub-lesion injections of Bleomycin or Interferon.

As a surgeon, I personally prefer to excise them under local or regional anesthesia and cauterize their roots with bovie. This rids the patient of them in one session.

65

DANNY'S STORY

Danny was the son of Diane C., my yoga teacher. Her classes consisted of muscle stretching and body balancing followed by periods of relaxation and meditation—they were very therapeutic.

One day, she brought me her ten-year-old son with tears in his eyes and his hands wrapped in bandages. He said, "I am embarrassed to have my classmates see my hands and be disgusted to touch me or even anything that I have touched. I'm even disgusted to eat with my own hands." While he was sobbing, his mother removed the bandages, revealing a large number of ugly warts covering his fingers, palms and dorsum of the hands. They looked like mushroom growths. They were very unsightly and justified his anguish. Danny had already tried creams, solutions, acids and cauterization but nothing helped; washing with soap simply resulted in a burning sensation. They had consulted many dermatologists with very little improvement. His mother wanted something done drastic, there and then.

Figure 8: Multiple warts on hands which were unsightly.

THE OPERATION

The patient was sedated with 10 mg. of Valium IV. I prepped his right arm from elbow to fingers and "Boers" anesthesia was introduced by emptying the veins and applying a tourniquet and filling them with Xylocaine. The degree of sensation was examined with a needle prick and found to be adequate. The warts were excised one by one from their roots by scalpel, scissors or bovie, and the roots were cauterized. There was no bleeding because of the tourniquet effect. After releasing the tourniquet, the roots of the warts that were oozing were re-cauterized with a bovie, resulting in the stoppage of any bleeding. Vaseline dressings were applied.

About 35 to 40 warts were removed from his right hand in this session. A pressure dressing was then added to the hand, with the arm kept in a sling. No antibiotics were given post-operatively and a Fentanyl patch of 50 mg. was applied to his chest. The patient was told not to touch the dressings on either hand until he came the next week to have the same treatment to the left hand.

THE MIRACULOUS EVENT

When he came the following week, he was very apprehensive about the appearance of the operated hand and the anticipated surgery on the other hand. When the dressing on the treated hand was removed, the hand and fingers were a little swollen and some of the scars were covered with scabs but not at all unsightly. He was pleased with the new, almost normal, appearance of his operated right hand and wanted the same surgery on the left hand.

We had everything prepared for the surgery, but when the covering bandages were removed, we were all surprised to find the warts we saw the week before on the left hand had spontaneously disappeared and the hand appeared normal without even scarring or remnants.

I had no explanation as to what resulted in their dramatic, spontaneous cure. God's cure comes in various ways.

NOTE

As these warts disappeared spontaneously, I've seen tumors disappear and miraculous cures by spontaneous regression of cancer. It can happen to anyone by the grace of God. The only requirement is a belief and anticipation of miracles. Diane believed in miracles, prayer and meditation.

I believe that all cures come from God, sometimes by assigned people, inspired techniques or by His direct intervention. So, don't let anything scare you. Whatever happens, pray with hope and faith, and it will come to pass, and you will be guided or cured. (More on this subject in Part II of this book.)

SECTION II
SERIOUS VASCULAR CONDITIONS

DISEASES OF BLOOD VESSELS

INTRODUCTION

The most common disease of the blood vessels is arteriosclerosis. It's the biggest killer of patients in civilized countries. Arteriosclerosis is the result of the deposit of animal fat like cholesterol on or under the inner lining of the blood vessels (intima). These deposits progress and narrow the lumen of the blood vessels causing symptoms, depending on where they occur, and the degree of narrowing. If they occur in the vessels of the neck (carotid) they can cause transient ischemic attacks (temporary lapses in consciousness or speech). That could culminate in a stroke.

If arteriosclerosis occurs in the coronary arteries it can cause angina on effort. Total occlusion causes *myocardial infarction* (a serious heart attack), which could be fatal.

Moderate arteriosclerosis of the femoral arteries can cause *intermittent claudication*, (pain in the calves when walking.) Resting pains are an ominous sign of severe occlusion. If complete, it results in gangrene of the leg. Gangrene without infection is referred to as *dry gangrene*.

Diabetes makes the patient susceptible to infection, more so when there is arteriosclerosis and lack of circulation in that part of the leg. "Diabetic gangrene" can be treated only by amputation to save the patient's life from bacteremia or septicemia which can result from a "diabetic foot."

73

In the early stages, diabetic infection in the leg can be treated with intravenous antibiotics and control of the sugar in the blood. If occlusion occurs on top of the infection, that part of the leg becomes dead and needs to be amputated.

There are various risk factors which can predispose a patient to arteriosclerosis. They are:

1. **Family history** indicating hereditary weakness or narrowing of blood vessels from birth. When this occurs in the heart, some members of the same family will die at a young age. Hardly any treatment that can change that at the present time.

2. **High cholesterol**. This results from a diet rich in animal fat, e.g., butter, fatty meats, some milk products, etc. Some people cannot metabolize even small amounts of animal fats. Such patients will need medications. With a blood test we look for the total cholesterol and the LDL (low density cholesterol). The last one is the most dangerous and is referred to as "bad cholesterol." Total cholesterol should be below 200 mg and the LDL below 100 mg.

Another element which can cause occlusion in the blood vessels is called *homocysteine*, an amino acid which if not metabolized and dissolved can also cause occlusion in the blood vessels. But it can be easily treated with Folic Acid and Vitamin B12 which help in dissolving it.

Lipids can also occlude blood vessels. These are fats which are not of the animal source, but if consumed in big quantities can have the same effect as cholesterol. Lipid count should be below 400 mg.

3. **Cigarette smoking**. Nicotine causes constriction of the blood vessels and chronic damage to the heart muscles. This predisposes to heart attacks or heart failure.

4. **High Blood Pressure** can push the lipids and cholesterol under the intima and cause a plaque (an aggregation to cholesterol and fat). Because blood pressure is higher in the arteries than the veins we rarely see arteriosclerosis in the venous system. Blood cells can be crushed while passing over these irregular plaques and cause a clot that could be fatal. If hypertension is not controlled, it can dislocate a plaque, occlude the blood vessel and cause sudden death. That is why these plaques, when discovered need angioplasty or bypass surgery, followed by strict diet so that they don't recur in other places of the blood vessel.

5. **Diabetes** can predispose to infection; obesity predisposes to diabetes.

6. **Obesity** is a risk factor because of accumulation of excess fat around the blood vessels of the heart.

7. **Stress** is a very serious factor in producing a heart attack. It can result in a sudden rise in the blood pressure and put the patient in a defensive adrenergic state. Adrenalin constricts the blood vessels and is the straw that breaks the camel's back. Many patients with controlled coronary artery disease, when exposed to sudden stress, anger or conflict with another person may get a sudden, fatal heart attack.

Stress should be avoided and treated by spirituality and faith. Its dangers should be explained to the patient, and the patient should convince himself that there is nothing in this world more important than his health.

Leaving problems to a higher power rather than ourselves gives us peace and tranquility. As stated in the Bible, "Peace I leave with you. My peace I give to you, not as the world giveth will I give unto you. Let your hearts not be burdened nor be dismayed."

DIABETIC GANGRENE

As I alluded to in the introduction, diabetic gangrene is a combination of arteriosclerosis and infection resulting in putrefaction of part of the leg which need to be amputated to avoid its spread to another part of the leg or death from septicemia. It usually begins at one of the toes and travels up the leg to the thigh. Experience proves that lower amputations often fail and most of the vascular surgeons prefer to go straight to above-the-knee amputations where circulation is better and thus avoids repeated failed lower ones.

In my opinion, the reason lower amputations fail is lack of circulation in the flaps of skin that cover the site of amputation. They eventually necrose, especially if closed with tight sutures. Additionally, closing the flaps prevents the infection from going out and forces it to ascend up the leg, necessitating higher amputations.

As you will see in the following story, again I had to deviate from the usual and customary to save my aunt's leg.

CLOSE TO HOME: MY AUNT SIMHA'S STORY

My aunt was fifty-years-old when diagnosed with gangrene in one of her big toes. She was overweight and diabetic with poor circulation in her legs. Although she never smoked, her husband was a three pack a day smoker, so she was exposed to a lot of secondhand smoke.

The most famous vascular surgeon in the city was consulted, and he recommended an above-the-knee amputation of that leg. He explained to the family that if we started by amputating a toe, the gangrene would travel to the lower leg due to lack of circulation as a result of infection. Then he added, "In this situation, if amputations at this level (below the knee) fail, then we have to go above the knee where the circulation is more adequate. As a surgeon, I have seen this happen again and again." I appreciated his honest opinion.

I appreciated the honesty, but was saddened by the prediction. I knew that she was obese and too weak to use crutches for ambulation. If restricted to bed, she would develop bedsores and her diabetes would become more severe due to a lack of motion. Rehabilitation would be hard, if not impossible.

This was my aunt whom I loved very much. I kept on thinking, over and over, that *there must be another way* when actually there wasn't. I thought and prayed and asked for guidance or inspiration. Suddenly a thought came to me: Why do lower amputations fail? The flaps do not survive the lack of circulation and the infection travels up because the sutures block its exit. What if we leave the flaps open and save them from the pressure of the sutures and let the infection go out instead of going up in the leg? This was not the conventional way, but just a hunch. I suggested this to my aunt and cousins and they were all for it.

Rather than admitting my aunt to the University Hospital where they would amputate her leg, I took her to a suburban hospital for same day surgery. I amputated the toe and left the flaps open, packed the space with Iodoform, and dressed the foot with loose bandages.

A.   B.

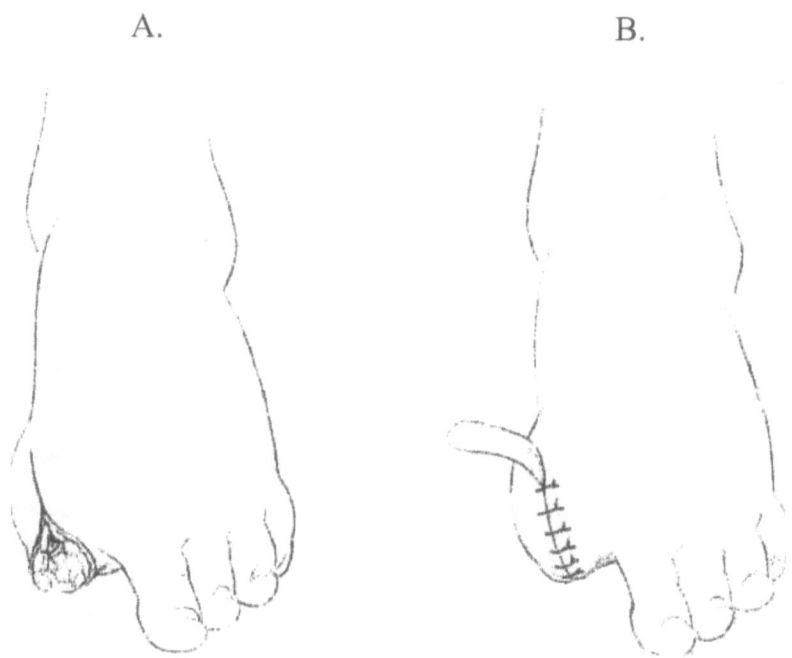

Amputation of a big toe
  A. On the left, keeping the flaps open.

  B. On the right, suturing the stump after amputation

Figure 9

With the help of a visiting nurse, we were able to control the diabetes, change the dressings and teach my aunt how to do it herself.

In about ten days, the pain subsided, the flaps remained viable, the infection was controlled and we were able to wean her from the antibiotics. In four weeks, the flaps approximated themselves and gradually stopped draining and the wound closed itself. The surgery was successful and my aunt could use her two legs for the rest of her life. Ambulation is very important for survival.

DISCOVERY

The reason for failure of amputations is suture closure. The stitches cut off the vascularity of the flaps especially when the flaps post-operatively swell. In diabetic gangrene, closing the flaps closes the outlet for the infection, and it travels upward. Since that experience, I made it a point to leave all amputations, at any level, with adequate open flaps. Eventually, they would heal around the muscles and close by themselves. Changing the dressings more frequently is, by far, better than re-amputation. Again, such findings are inspired by love for my patient and profession. The deep desire to give our patients a better life and not condemn them to a possible cruel fate is the motivating force. I hope other doctors will do likewise. I think we should not only save their lives but try our best to give them the best possible quality of life.

CLOTTED BYPASS GRAFTS

INTRODUCTION

If a blood vessel is occluded by arteriosclerosis, the occlusion should be bypassed to avoid distal ischemia (lack of blood supply), and if the blood vessel dilates and develops an aneurysm that may rupture and kill the patient from hemorrhage; the aneurysm should be excised and replaced by some kind of conduit to keep the circulation distally.

My experience with replacing a blood vessel dates back to 1957 when I was an intern under Professor Taylor and Mr. McGowen. Mr. McGowen asked me to be on call for someone who was brought in dead, and to harvest both his femoral arteries from groin to knee and put them in a preservative solution. They were going to use them to replace two syphilitic aneurysms in the backs of the legs (the popliteal area). One weekend an electrician was hit by a very high voltage current and died instantly. The ER doctor called me and I removed his femoral popliteal arteries to use them to bypass the resected aneurysms. Such a procedure was new at that time.

A few days later, the patient was brought into the operating room, put to sleep and turned on his belly. The aneurysm was dissected and resected through the popliteal fossa, and the arteries I had harvested from the dead electrician were used to bridge the gap. The operation was performed by Mr. McGowen, guided by Professor

Taylor, and naturally, I was used as an assistant. A week later, the same procedure was performed on the other leg. The surgery on either side took about six hours and three units of blood, which I collected from donors and cross-matched, as there were no blood banks at that time.

Fifteen years later, when I myself became a vascular surgeon, I found it was best to make these operations from the medial side of the leg rather than the posterior approach which my mentors used at that time.

When I came to Texas in the late 1960's, I was told that Dr. Cooley and Dr. Michael DeBakey were using cadaver aortas to replace abdominal aortic aneurysms. One day when they were faced with a ruptured aneurysm and there was no cadaver available, it was said that Mike asked his OR supervisor to go to his office and cut a piece from the back of his nylon shirt and sew it in the form of a 3 cm. diameter tube and autoclave it. It is said he successfully replaced the bleeding aneurysm with the nylon graft, opening the door for the use of artificial grafts to replace blood vessels. This daring act paved the way for experimentation using other materials like Dacron and Teflon, some porous and some nonporous. This research even led to the use of fetal umbilical cords which were thought to be least rejected by the body and less likely to thrombose. Dr. Cooley experimented with various grafts and that led to the best material and porosity that now carry his name.

In spite of all these improvements, artificial grafts have a limited longevity. They can be used as temporary bridges, and if the patient is lucky, he might develop his own collateral circulation around the graft. The reason the grafts clot is either a progression of the arterial slowing of the flow of blood inside the graft (stasis), and that leads to clotting of the blood. That is why trying to disobliterate the graft by embolectomy or proteolytic enzymes usually fails, and a new graft with a new inflow source and patent outflow segment to allow the blood to flow freely to supply the limb, can save it.

PERSEVERANCE PAYS: BERNICE'S STORY

Bernice was a sixty-year-old white female. She came to me with a right superficial femoral artery occlusion that caused leg pain at the slightest movement. That necessitated a femoral-popliteal bypass for which I used a Dacron graft. This graft lasted about five years and then became occluded, resulting in recurrence of ischemia in her right leg. The femoral graft was removed and replaced by an ilio popliteal umbilical vein graft. This lasted three years and then thrombosed, causing recurrent severe ischemia to the right leg. The radiologist suggested using thrombolytic solutions to disobliterate the graft. Predicting the difficulty of the operation, I agreed to let them give the thrombolytic therapy a chance. After five days of continuous IV infusion with Thrombokinase, the graft remained occluded and the radiologist informed the patient and me that she needed an above the knee amputation. Bernice was a very emotional patient and objected severely to such a radical procedure. At the same time, her general condition did not allow an abdominal exploration. Hence, I planned a left to right femoral artery bypass. During this operation, the left femoral artery was patent. A Cooley 8mm Dacron graft was hooked to the left femoral artery in the usual manner after heparinization, and tunneled through the lower pelvic area to the right side.

On exploring the right femoral artery, I found it to be totally calcified with no lumen. I tried to dissect the deep femoral artery up to the middle of the thigh but could not find any open lumen to hook to the new Dacron graft. At that moment, I thought the radiologists' conclusion that there was no choice but amputation was correct. Based on a faint idea that I might locate a branch of the deep femoral artery, I continued to dissect for almost two hours under total discouragement by my assistant—a board certified vascular surgeon—and the anesthesiologist. Stopping for a moment to ask for guidance, I continued further dissection, and lo and behold, I found a small patent tributary. I immediately tailored the graft to the small diameter of its lumen, and using magnification, and the experience from coronary artery surgery, I succeeded in establishing an anastomosis.

The chances of success were slim, but post-operatively the staff at the ICU and I were gratified to feel a faint anterior tibial (artery in front of the foot) pulse confirmed by a Doppler sound.

That night, I still remember, I called the ICU nurses every hour to ask if they could hear Doppler pulses. The responses were variable; yes, no, maybe, but they assured me that the foot felt warm, which means viable.

Early in the morning, I went to visit the patient. It was a very dramatic moment when the patient asked whether she still had her leg or if it had been amputated. I told her we should remove the bed covers and find out. The pleasure in her eyes was indescribable when she sat up and saw her two feet in place. With difficulty, I could hear a very faint Doppler pulse between the first and second metatarsals.

Although, statistically the results of such a procedure were not promising, I prayed to God not to disappoint her. She was seventy-seven at that time. I gave her Plavix and Vasotec, and then referred her to physical therapy for ambulation. She was discharged ten days later, still able to use both legs.

On her 80th birthday her son, who was a developer, built her a home next to his in Rockford, Illinois. He threw a surprise birthday party for her and invited us. My office staff, my wife and I drove eighty five miles on that weekend to celebrate with her and her family. She jumped with joy when she saw us and it was very gratifying when she walked us around her house on her own two legs. She was able to enjoy ambulation even up to the last day of her life thanks to the extra two hours of dissection during her operation.

I have no doubt that success in this case was due to my love and compassion, perseverance and prayer, and the supportive post-operative care of her sons.

THE PLACE OF COMPLEMENTARY MEDICINE IN MEDICAL PRACTICE

INTRODUCTION

In my office building in Chicago, I had an auditorium where I gave a monthly lecture to my patients, their friends, and the public, helping to further their medical education.

At the end of one of those lectures, a patient asked me about *free radicals*. At that time I wasn't knowledgeable on the topic but I promised the attendees I'd study the subject and report my findings. I not only researched the subject of *free radicals* but also studied their treatment using antioxidants.

In another lecture about peripheral vascular disease, after listening to the risk factors and modern treatment with grafts and so on, one of the attendees asked me if I used *chelation therapy* in vascular disease. I told him at that time I knew about its use in lead poisoning and other heavy metal toxicity, but didn't know about its use in vascular disease. Nonetheless, I again promised the listeners I would study the subject and report back to them at a later lecture. I looked it up on the internet where I found a lot of testimonials from people with intermittent claudication who felt cured by that therapy. A further search to find more information led me to the "American College for Advancement in Medicine" where they offer courses on chelation therapy and other alternate medicines at their meetings in different cities in the country. The next one was scheduled to be held in Phoenix, Arizona, so I registered myself for that four-day conference and attended two courses on the subject. The first course for the MD's and DO's focused on its history and the scientific basis of its use. The second course taught our medical technicians how to prepare it and how to use it, and what to watch for during a treatment. In the exhibits, I was surprised to find hundreds of books about this subject, some for doctors and others for patients. What impressed me at that meeting was that many MD's who came for a refresher course said their patients benefited from chelation therapy and the safety factor of the procedure had zero mortality. Since neither Medicare nor regular insurance paid for this treatment, the patient would either pay a small fee in cash or doctors did it free for the benefit of their patients.

The other aspect was the thousands of patients who left their names and phone numbers to bear witness about its benefits. My medical technician took the appropriate course with me. After I read a number of books about the subject, and since it did no harm, I decided to evaluate it for myself when the occasion arose or when a patient demanded it.

A month later, just such an opportunity arose when a patient came asking for it. Here is his story.

REJECTS SURGERY—IS THERE AN ALTERNATIVE? RICHARD'S STORY

Five years after Bernice H. was discharged her sixty-year-old son, Richard H., came to see me as a patient. He told me he had "inherited his mother's disease." When I asked him what he meant by that, he said he could not walk a block without his calves becoming painful and tired. I performed a Doppler study on him which revealed a good femoral pulse but weaker popliteal and very faint pedal pulses. Arteriograms revealed bilateral superficial femoral artery occlusions but patent deep femorals.

He also told me that he had seen how his mother suffered from various grafts and asked me if I was equipped to give chelation therapy. He had heard of its benefits from friends with similar conditions and wanted to try it before considering surgery. He was happy when I told him that I was familiar with it, and asked me to try it on him. I agreed but added some conventional medicine (Plavix and Lipitor for his cholesterol–285–and the newly discovered Nitric Oxide therapy, a vasodilator).

For the next two weeks, I gave him IV chelation therapy twice a week. He reported an improvement in his claudication and so began a weekly regime.

After about sixteen therapy sessions, he stopped showing up. I was very anxious to know why he stopped coming. When I called him, he told me he suffered no more claudication and that, in fact, he was participating in a golf tournament but promised to show up when the tournament was over. When he came back, Doppler studies revealed a good pedal pulse. He agreed to come once a month and then once every two months, and then once every three months. I tried to repeat the angiograms on him, but he refused and questioned the necessity for them since he recovered totally from the claudication.

REGARDING CARE FOR INTERMITTENT CLAUDICATION AND CHELATION THERAPY

The theory behind chelation therapy is that it dissolves the calcium in the plaque (calcium being a heavy metal) and turns a hard calcified plaque to a soft non-obstructive one. Over the years, I tried this therapy on other patients for claudication and even mild angina with improvement in the symptoms. In one patient, I proved this by having my cardiologist repeat his angiograms six months later. This patient had diffuse disease and tiny blood vessels and was inoperable. The post treatment angiogram revealed an increase in the length of the left coronary vessels from the mid heart to the apex and the right coronary artery, which was totally occluded, revealing a lumen of 20%. Together with this, I gave him cholesterol lowering drugs, calcium channel blockers for hypertension, (the patient could not receive beta blockers because he had emphysema), as well as Plavix. I also asked him to stop smoking.

I do not think that this is an alternative to bypass surgery or angioplasty, but it could be of use in patients with diffuse, inoperable disease. I also notice that chelation energizes patients and has some benefits in "chronic fatigue syndrome."

Many patients are accusing doctors of arrogance for not discussing alternatives with them and that alternative techniques are even described as quackery. Such answers could hurt patients and make them lose trust. Doctors should be more open-minded and have the humility to accept new ideas.

Today, alternative and complimentary medicines are experiencing revitalization. New developments include a growing professionalization, the creation of several complimentary medicine journals, the establishment of the Office of Alternative Medicine at the National Institute of Health, the proliferation of books and conferences discussing research, clinical applications, data survey, and even alternative medicine schools. Americans spend billions of dollars per year on alternative treatment options.

THE CHRONICALLY SWOLLEN PAINFUL LEG

INTRODUCTION

The causes of a swollen leg are many:

- Deep vein thrombosis obstructs the return of venous blood to the heart.

- Congestive heart failure.

- Inflammation or infection, such as cellulitis.

- Arthritis.

- Lymph edema as in Elephantiasis.

- Lymph node obstruction, resulting from radical lymph node dissection or it can be congenital. Iatrogenic, e.g. ligation of a major vein during surgery to stop a massive hemorrhage.

SWOLLEN, PAINFUL LEG: PAULINE'S STORY

This seventy-year-old Caucasian lady was referred to me in Martha Washington Hospital because of a twelve-year persistent, painful swelling in her left leg. During that period she was seen by several doctors giving her various therapies, including Heparin, Coumadin, Plavix, thrombolytic agents, as well as antibiotics. Nothing helped, and she was desperate. Her condition was pathetic from the consistent pain and swelling in her left leg. She suffered severe venous varicosities on the left side of the abdomen as well. This went on for twelve years.

When I saw her in consultation, she gave a history of a surgical pelvic procedure, which resulted in massive hemorrhage. She had been told that a major vein was ligated to stop the bleeding. A venous angiogram revealed a total occlusion of her left iliac vein with a huge collateral circulation and varicosities formed in the body's attempt to bypass the obstruction.

This was a rare case, and when I showed her venograms in our vascular society meeting later, many vascular surgeons made the diagnosis of arterio-venous fistula due to the appearance of severe varicosities around the occlusion. AV (arterio-venous) fistulas have a similar appearance on x-rays; however, there was no bruit or cardiac decompensation. This ruled out any diagnosis of a fistula.

I had never encountered such a case before. I scratched my head and asked for inspiration for a solution to her obstructed left leg venous circulation, and reviewed the literature on vein thrombosis. At first, I thought about a femoral vein to vein artificial graft to take the circulation off the obstructed left leg to the right open leg circulation. This procedure is used for arterial occlusions where the blood flow is forceful but not in venous where the blood flow is comparatively sluggish. Thus, I thought of using her own vessels for a bypass, and further thinking led me to using her short saphenous vein (this is the same vein we harvest for coronary bypass surgery) instead of a synthetic graft.

Under spinal anesthesia I dissected her left short (deep) saphenous vein, tunneled it subcutaneously to the right side, and anastomosed it to the right femoral vein, already exposed. With that, the obstructed circulation in the left leg could go to the open right iliac vein. In this case the short saphenous vein takes the circulation in the same direction, i.e. upwards, as always and thus does not need to be reversed (When used in coronary bypass it is reversed to carry the circulation from the aorta above to the coronary artery below).

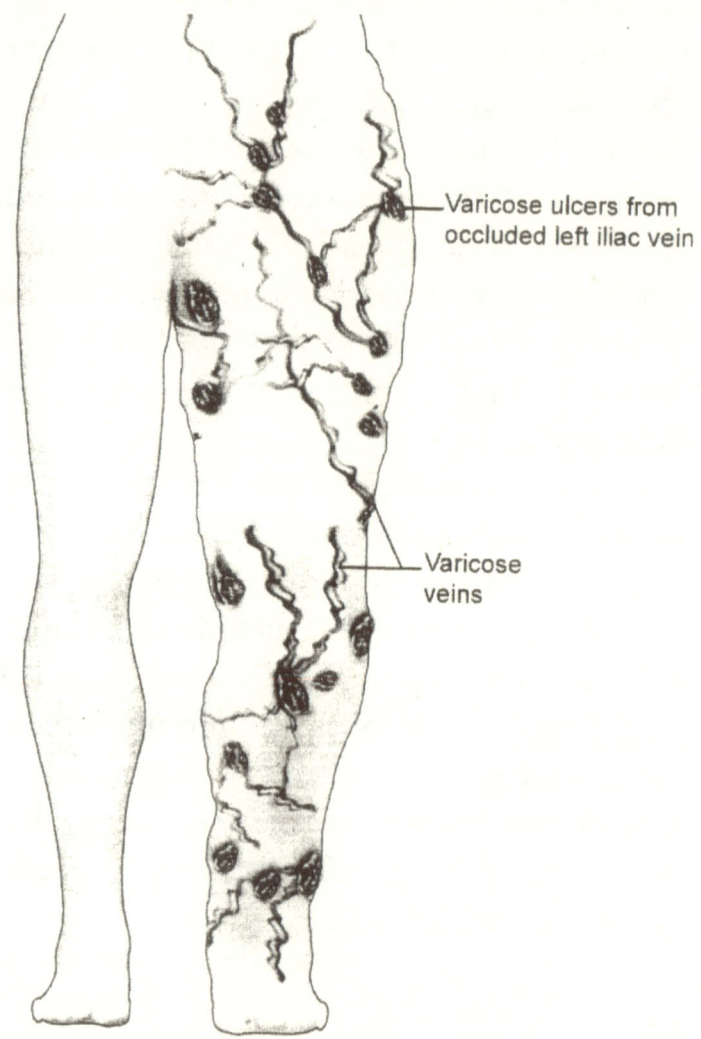

Varicose ulcers from
occluded left iliac vein

Varicose
veins

Swollen ulcerated painful left leg

Figure 10

Immediately after the operation, the varicosities in her left leg started to disappear, and within a week the swelling completely vanished. And this after twelve years of suffering with her left leg swollen and elevated on pillows with no improvement. You cannot imagine her happiness and mine when I discharged her from the hospital with two equal legs, no pain, and no ugly varicosities on the surface of the left leg and pelvis.

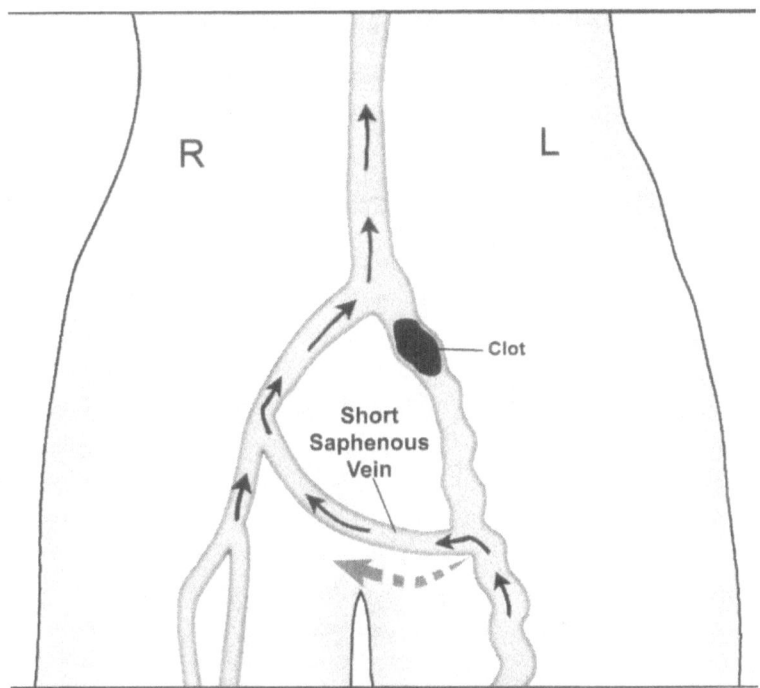

Figure 11: The operation.

NOTE

The reason for success in this case was inspiration and a creative imagination which allowed me to visualize the effect of using an extra open vein in the left leg, (the short saphenous vein), and transfer it to the right side which is patent. Femoral artery to femoral artery bypasses are usually used in arterial occlusions, but, this was the first time I was inspired to use a vein to vein bypass using her veins.

It brought tears to my eyes when the patient said to me on her day of discharge relieved from her suffering,

"Sometimes, to the world you might be just one person, but to me, you are the world."

The greatest reward in a surgeon's career is when he can make a difference in another person's life.

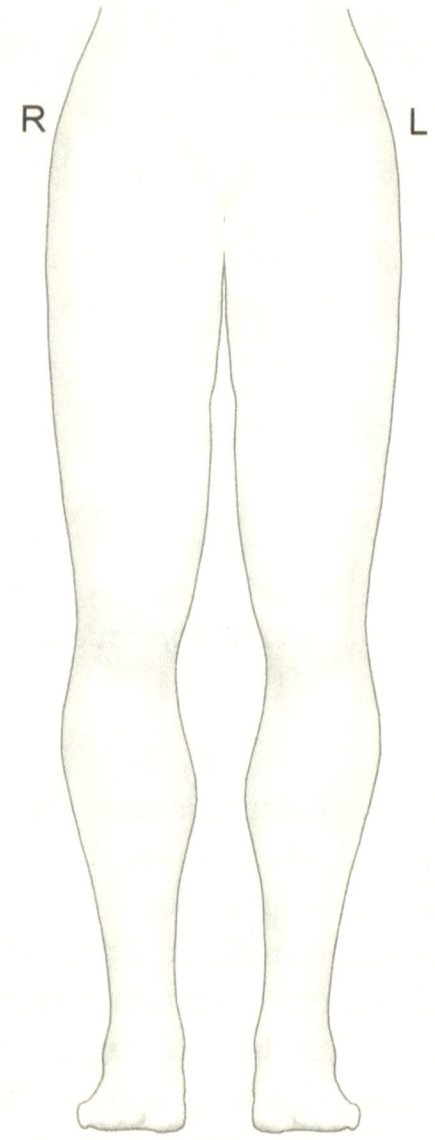

Figure 12: Both legs after repair.

ARTERIOVENOUS FISTULAS

INTRODUCTION

Arterio-venous fistulas are connections between an artery and a vein resulting from traumas such as stab or gunshot wounds. The high pressure arterial blood goes into the low pressure vein increasing the venous return to the heart and resulting in heart failure.

The treatment consists of separating the artery from the vein and repairing the hole in both of them. The unique thing about the case that follows is that the fistula was in an inaccessible place at the base of the neck and under the manubrium. No description of such a case was yet to be found in literature and so many surgeons were consulted by the family physician and the patient. They were told that this fistula was out of reach and inoperable.

I spent weeks trying to plan an operation. It was a great challenge. I did not want to try and fail which would cause more distress to the patient. After weeks of planning, I was inspired with a reasonable plan, but it involved opening the chest, splitting part of the sternum and exploring the axilla.

You should never give up on difficult patients. If you love them and are determined to cure them and pray to find a way, you'll be inspired to find a cure. Inspiration is one of the basics for creativity. It was used in solving problems for patients considered incurable, hence considered miraculous.

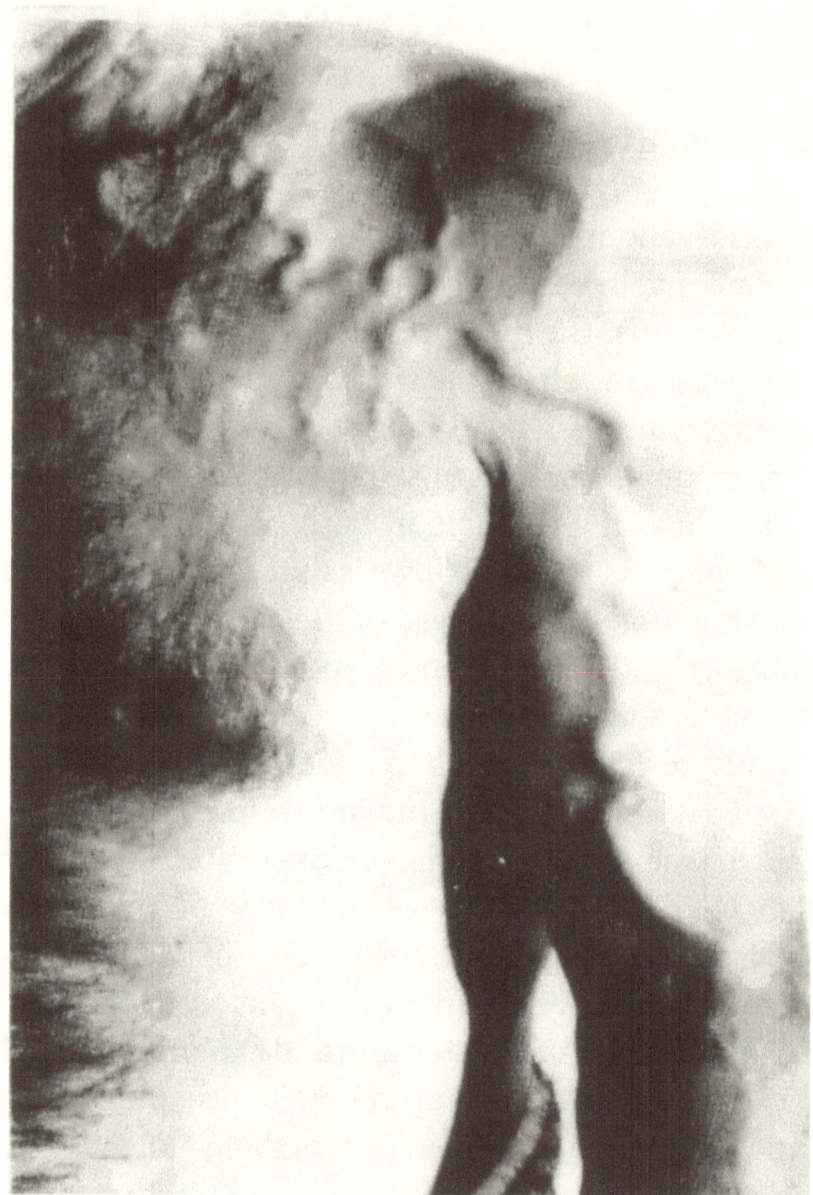

Figure 13

A FISTULA BEYOND REACH: CARLOS' STORY

The patient was a forty-seven-year-old Puerto Rican male. At a young age he was shot in the neck. The bullet went through the base of the neck at the region of the sterno-clavicular junction to the top of his left shoulder. After resuscitating him, he could feel a thrill and hear a murmur coming from his lower neck to the upper chest. He was diagnosed with a deep subclavian arterio-venous fistula.

In consultation with various surgeons he was told that his fistula was out of reach for repair. The arterial blood with its high pressure went into the low-pressure veins in his chest, causing snake-like varicosities, extending from the left side of his neck to below his rib cage and along his left arm. He gradually became more and more fatigued and short winded. He also developed palpitations. In spite of medical treatment, he also lost strength in his left arm. Arteriograms and cardiac catheterization confirmed the diagnosis and revealed that his heart was enlarged. (See photos with his permission on page 103.)

My thoughts raced. Should I turn him down for surgery, as all other surgeons did, or use my creative imagination to save this man from deteriorating further? What if I meet with uncontrollable hemorrhage or the operation fails?

I researched many books and atlases for pictures of blood vessels in the neck and drew a plan in my mind. I decided that without risks there would be no success, and with no action on my part the patient's heart would fail and his life would be compromised. The referring physician was pleased when I told him that I had a plan and already scheduled the patient for surgery.

The five-hour surgery was not easy. There are always surprises in the operating room. When I finally got to the area of the fistula, I was faced with an aneurysm in the subclavian artery. It was pressing on the brachial plexus on one side and covering the subclavian vein on the other side. Finally I was able to dissect the aneurysm from the brachial plexus, get control with Cooley clamps on the proximal and distal subclavian artery, resect the aneurysm and lift it up where I could see the subclavian vein (the vein was similarly controlled proximal and distal to the aneurysm).

Both artery and vein were repaired by bridging grafts. Both the patient and I were gratified in spite of the big incisions. The distal control of the blood vessels was in the sub clavicular area as an extension of the axillary artery. The proximal control was more difficult because the aneurysm was intra-thoracic. It was approached by a hemi-sternotomy, plus an incision in the third intercostal space. Raising a flap of the left upper hemi-thorax to get access to the subclavian vessels distal to their branches allowed complete excision of the fistula and repair of both artery and vein in continuity.

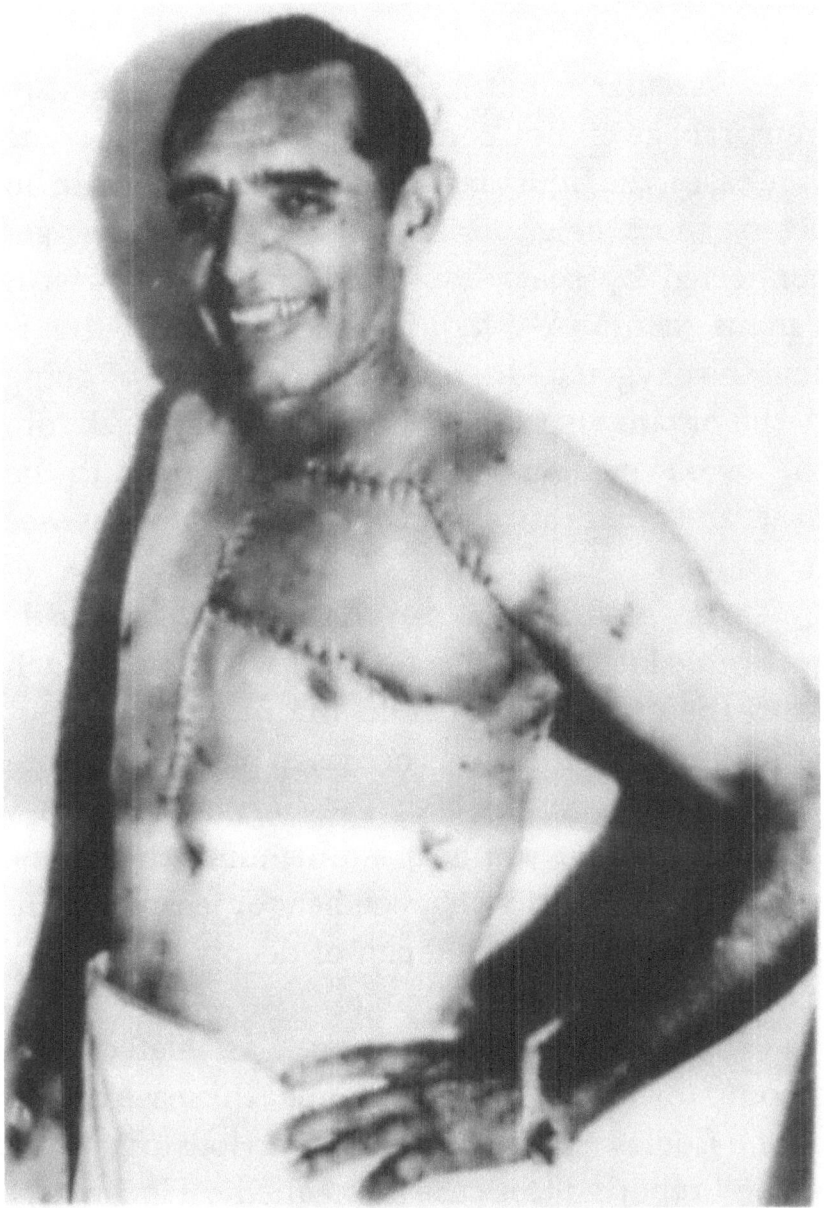

Figure 14: Post-op picture, published with patient's consent.

NOTE

Vascular injuries can cause a severe hemorrhage if not controlled promptly. In some rare cases, the injury can result in facing holes in an artery and a neighboring vein. The two holes get connected by scar tissue and create the arterio-venous fistula (A-V fistula). The danger of this is that the oxygenated arterial blood, instead of going to the organs that it supplies, is diverted back into the venous system which returns the blood to the heart and causes an overload to it which predisposes it to failure.

An A-V fistula can happen in an easily accessible location (e.g. a femoral A-V fistula), such as in the groin or higher in the iliac vessels. But sometimes the A-V fistula occurs in places which are deep and difficult to access. The diagnosis of an A-V fistula is made if you hear a murmur at the site of the injury, similar to what you hear or feel when you put your hand on a cat. Repair of deeply located A-V fistulas can be a real challenge.

The referring physician was elated and accompanied me when I presented this case in one of the Cooley Society Meetings in Houston, Texas. When I reported this case in the medical journals, I received requests for reprints even from countries overseas, including the Eastern Block.

The real pleasure comes from not only having success in curing this "incurable" patient, but also in motivating my fellow surgeons to do likewise by being courageous, inspired, and creative when met with unexpected surgical challenges.

The spirit of being courageous, innovative and creative in the direst of medical circumstances was cultivated in me by my mentors. They invented great procedures and had the courage to implement them. Whenever I remember this story, I always send good wishes to them. I thank God for giving me the opportunity to follow in their footsteps and do something innovative when others thought it was impossible.

SECTION III

CARDIAC EMERGENCIES

CARDIAC EMERGENCIES

INTRODUCTION

In no other type of emergency surgery is there such a need for serenity, planning, and alerting the team and motivating them, as in this kind of surgery. It is dramatic, risky, and requires the support of a competent cardiologist and anesthesiologist in a hospital that is equipped to handle such emergencies. Open heart surgery is obviously much safer when done in a non-emergency situation. The mortality of coronary bypass when done prophylactically is down to one or two percent nowadays, and rises to between four and five percent if done during an acute myocardial infarction.

HEART VALVE DISORDERS

When the mitral valve is closed, the oxygenated blood cannot pass in the left ventricle and is returned back to the lungs which become congested and end in pulmonary edema. Then the whole body is deprived of oxygen.

The usual cause for mitral valve disease is almost always rheumatic fever as a complication of streptococcal tonsillitis. This was common in the 1940s and 1950s, before the antibiotic era. With the advent of antibiotics, the incidence of rheumatic fever in the United States and Europe has decreased considerably, but in the rest of the world where antibiotics are not

109

available, rheumatic fever and mitral valve disease are still common. At the present time, most of our valve surgery comes to us from abroad, mainly South America.

TWIN PREGNANCY WITH MITRAL INSUFFICIENCY

INTRODUCTION

Mitral regurgitation occurs when cusps of the mitral valve shrink and leave the mitral valve open all the time so that most of the blood coming to the left atrium is pumped back through the open mitral valve and returns to the lungs, causing pulmonary congestion, pulmonary edema, and heart failure.

In a pregnant woman, the stress of delivery can exaggerate the symptoms and lead to death, especially if the patient is pregnant with twins.

MARIA'S STORY

The patient was a twenty-two-year-old Mexican woman about three months pregnant with twins and had severe mitral insufficiency. She was referred to me by her obstetrician in 1974 for Mitral Valve repair for fear that she might go into cardiac failure during the delivery.

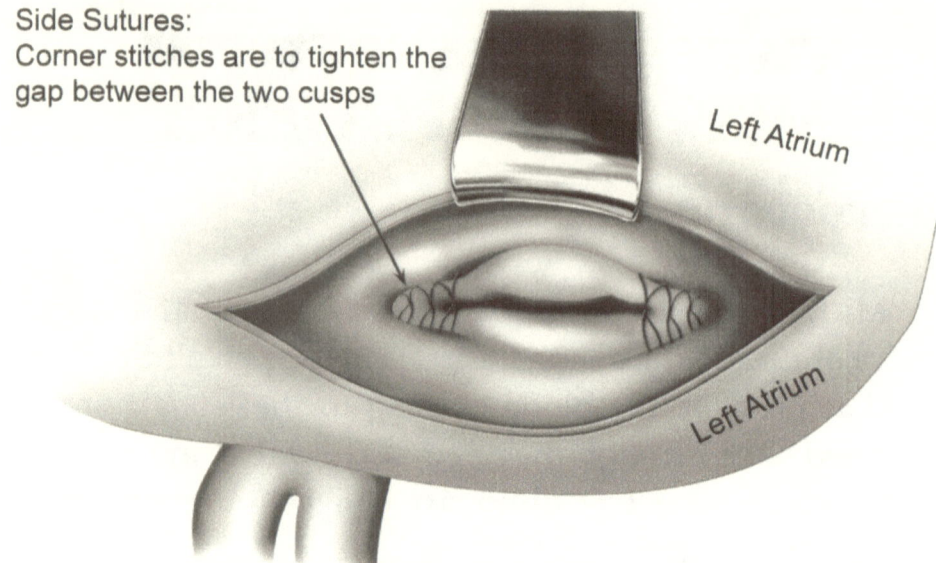

Side Sutures:
Corner stitches are to tighten the
gap between the two cusps

Left Atrium

Left Atrium

**Figure 15: Repair of mitral insufficiency by approximating the cusps
with side sutures.**

The patient was admitted to Mt. Sinai Hospital in
Chicago, Illinois and studied by the Chief of Cardiology
(at that time, Dr. Slotky). We decided to do a mitral valve
repair instead of replacing it with an artificial valve to
avoid anti-coagulation (which is needed for artificial
valves) during labor. The next day the patient was taken to
surgery and put on bypass, as usual. The left atrium was
opened. I found the mitral valve leaflets were soft and
pliable, but shrunken and curved inward, allowing a large
opening between them. Two figures of eight sutures were
put in the corners of the mitral valve commissures which
brought the leaflets closer together.

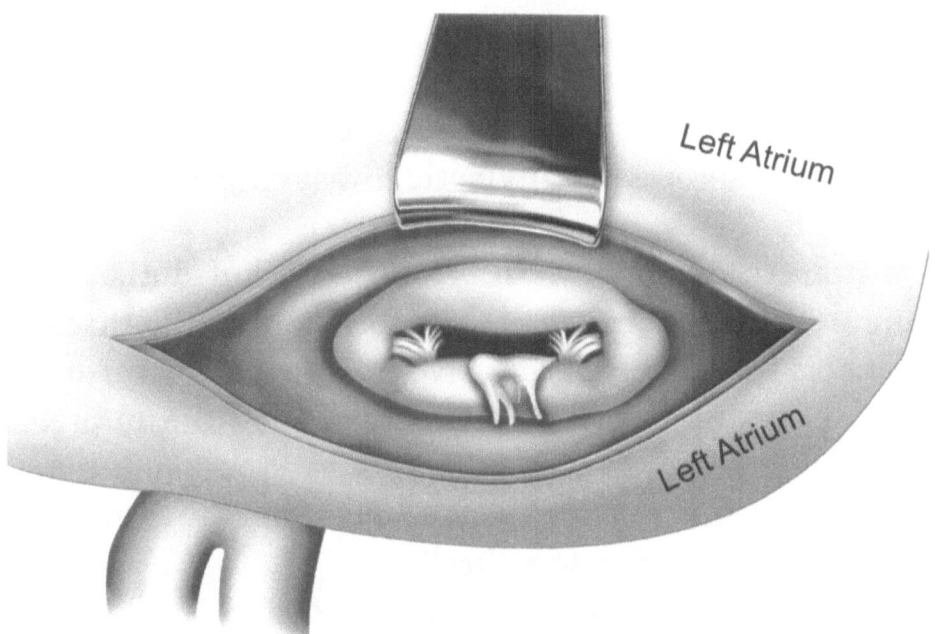

Figure 16: Ruptured chordae tendineae.

POST-OPERATIVELY

The patient did very well. The harsh systolic murmur, characteristic of mitral regurgitation, almost disappeared. On the third day, the patient was taken off oxygen by nasal cannula and allowed to ambulate. We decided to discharge her a week after the surgery. But on the day of discharge, I went to say goodbye, and was horribly surprised by the return of the murmur. There was a left ventricular heave (feeling a rising of the chest wall when you placed you hand on the left side of the chest), and it was clear the symptoms recurred and the patient was going into failure.

I almost had tears in my eyes. Why did a successful repair suddenly fail? I called my cardiologist, Dr. Slotky.

My impression was that one of the repair sutures at the corners of the valve had become loose. It's one of the most horrible disappointments for a cardiac surgeon to have a prophylactic, lifesaving surgery fail, especially in a young patient like her.

After studying her, Dr. Slotky concluded that she ruptured a *chordae tendineae*, which is one of the threads that hold the leaflets of the valve close to the myocardium. It's like an umbrella wire that does not allow the umbrella to open more than it needs. This unexpected rupture allowed the mitral valve cusps to open more than needed and the regurgitation to return.

The technique of repairing a thread-like chordae, and reattaching it back to a papillary muscle, was not perfected at that time in the United States, though a Dr. Carpentier in France was experimenting with it. It was possible further repairs in delicate cusps like hers could fail again.

The only alternative was to do a mitral valve replacement with an artificial valve. This would be a life saver under these conditions in spite of the nuisance of anti-coagulation. It was hard for her husband to understand, but he accepted the fact there was no alternative. Twenty-four hours later, the patient was taken back to surgery and the mitral valve was replaced with a low profile valve. This time, the patient improved dramatically and was discharged a week after the second operation on Coumadin.

Naturally, I followed the patient during her delivery. A week before the due date, she was put on "mini" heparin subcutaneously and had a normal delivery of two beautiful twin boys, delightfully replacing the prior crisis.

The rupture of her chordae was nobody's fault. It's a rare complication of *rheumatic carditis.*

To the best of my knowledge, the patient and her husband returned to Mexico a few months after the delivery. I felt an obligation to record it in this book in case it happens to another surgeon and patient.

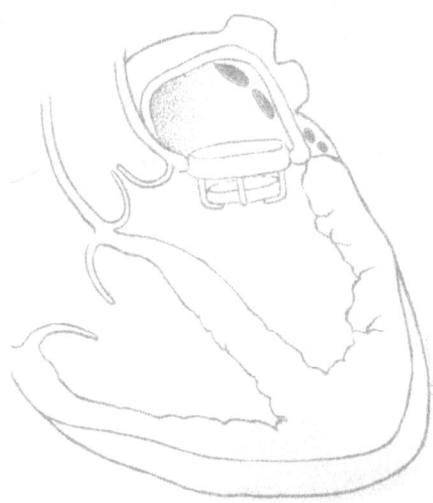

Figure 17: Mitral valve replacement. (Illustration adapted from Dr. Cooley's book, Techniques in Cardiac Surgery.)

NOTE

In retrospect, maybe it would have been wiser to replace the valve from the beginning. This is one of the hard decisions which surgeons and cardiologists face while trying to do the best for their patients. Artificial valves need anticoagulation postoperatively. We try to avoid mitral valve replacement during pregnancy because of the risk of anticoagulants during labor and delivery.

The miracle in this case is the development of open heart surgery and valve repair or replacement, without which this patient might have died during the delivery.

PREVIOUS EXPERIENCES

In the back of my mind, I recalled when I was an intern rotating in OB-GYN in 1958. I was called to help in the delivery of a woman with mitral insufficiency and twin pregnancy. When I arrived, a cardiologist was trying to resuscitate her from acute heart failure and a midwife trained to do vaginal deliveries, was present. Before delivering either of the twins, she died. When I reached my boss to inform him, he told me that there was an eight minute interval during which I could do a C-Section if the husband approved and save the babies. It was one of the most dramatic and traumatic experiences that I'll never forget, performing a C-Section on a dead woman and delivering two sons to the husband as he mourned his wife's death.

Without open heart surgery existing at that time, there was nothing else that could have been done to repair her valve and save her life. This experience came back to me when another young woman was referred to me.

Thanks to the progressive nature of medicine, twenty years later open heart surgery became available. Dr. Starr pioneered valve replacement, and repair became a viable alternative.

MITRAL STENOSIS

INTRODUCTION

The opposite of mitral regurgitation is mitral stenosis, also resulting from rheumatic fever. In stenosis, the cusps stick together leaving very little opening for the oxygenated blood to come from the lungs into the heart. That causes a backup of blood inside the lungs and ends in pulmonary congestion and, if not treated, in pulmonary edema whereby the patient drowns in their own fluids.

OLD AND NEW TECHNIQUES: ARLENE'S STORY

Arlene K. was referred to me by her family physician in a state of severe pulmonary congestion in spite of medications. She needed emergency heart surgery.

She was born in 1930 and between the ages of 8-10 suffered repeated attacks of streptococcal tonsillitis. This resulted in rheumatic fever which caused the two cusps of her mitral valve to stick together. At the age of twenty, she went into heart failure.

THE OLD SURGERY

In 1950, before the advent of open heart surgery, she underwent a *Closed Mitral Commissurotomy*. This consisted of putting a purse string suture around the left auricle, then making a large enough opening to introduce the surgeon's finger into the left atrium. He can thus feel the small opening between the mitral cusps and quickly pass it to both sides of this opening to break the adhesions in the cusps and create a larger space between them to allow the blood from the lungs to pass more freely into the left ventricle. This should be done quickly because when the finger is inside the opening it totally occludes it and no blood can pass into the heart, which could result in cardiac arrest. If it cannot be done quickly, the surgeon should do it at intervals allowing some blood to pass in between the finger fractures.

Once the opening is large enough, the surgeon

withdraws his finger and the purse string is tied to stop any bleeding from the hole that was created to admit the finger. The disadvantage of this method is that it is done blindly and could be incomplete. With time the adhesions might recur and cause re-stenosis.

This procedure kept her valve patent for 25 years. In 1975, at the age 45, the valve reclosed, and she was admitted to Mt. Sinai Hospital with symptoms of pulmonary edema. Her situation was considered an emergency, and she needed immediate surgery.

THE NEW SURGERY

In 1960, after years of research, John Gibbon of the University of Pennsylvania succeeded in inventing the heart lung machine. Circulation could be transferred to it, the heart emptied and opened, and the surgeon can work inside it under direct vision. The adhesions could be cut by scissors or knife, or if the valve is very deformed, one could replace it with an artificial valve. It also allowed Dr. A. Starr of Portland Oregon, with whom I worked in 1969, to replace a deformed heart valve with a mechanical heart valve for the first time in history.

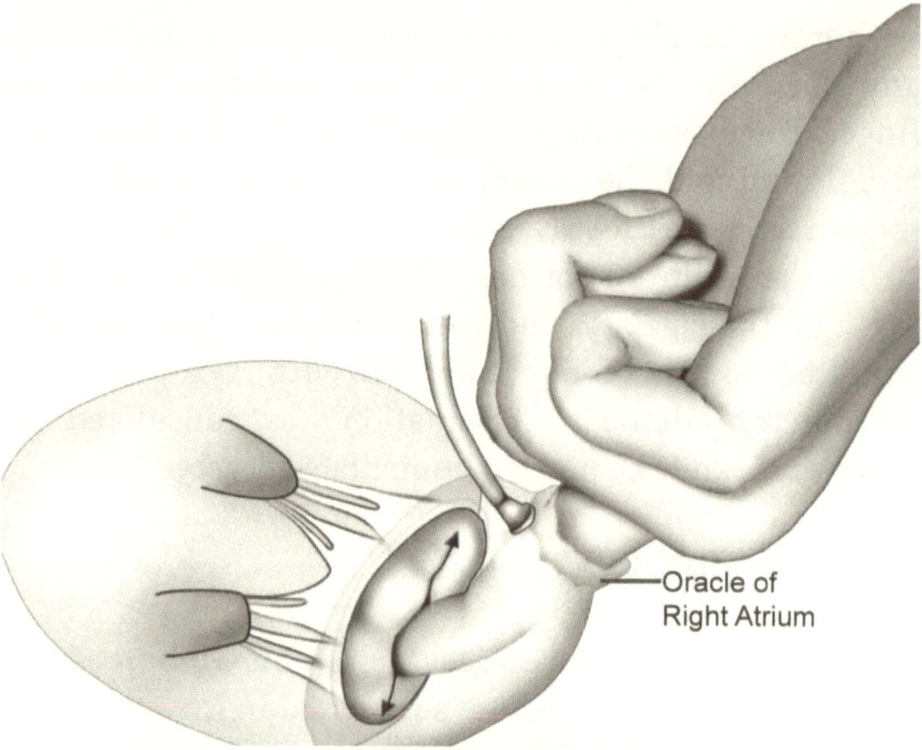

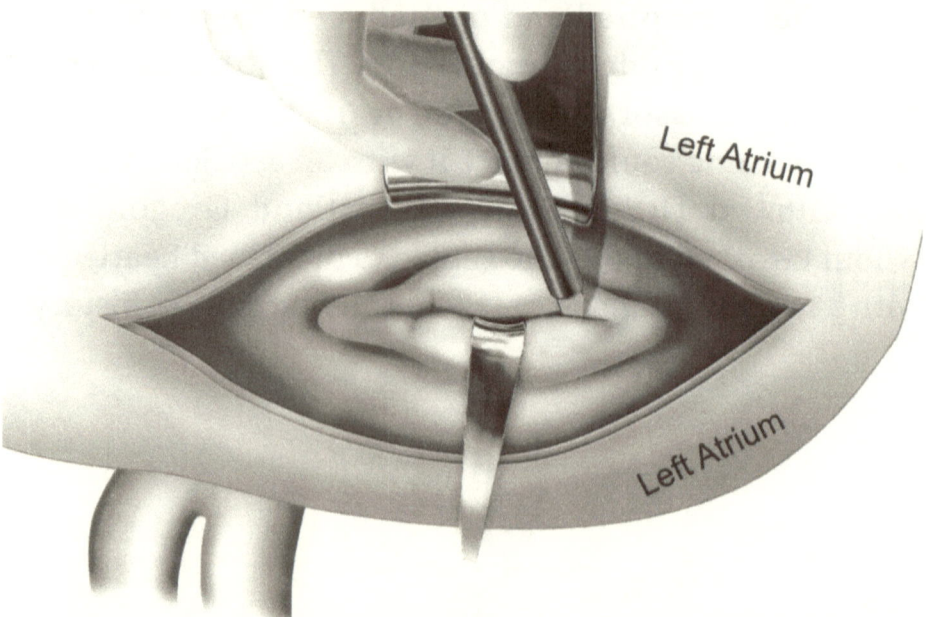

Oracle of
Right Atrium

Figure 18: Closed method—finger fracture. Old method.

Left Atrium

Left Atrium

Figure 19: New method—open mitral commissurotomy under vision.

120

THE OPERATION

When I saw Arlene, she was close to pulmonry edema, not responding to vigorous medical treatment by the cardiologist who already studied her. She was admitted to Mount Sinai Hospital for emergency surgery.

There was a debate as to whether to put her on an "aortic balloon pump" to decompress her heart, but I felt that the sooner we dealt with the cause of the problem (the stenosis) the better for dealing with the symptoms (congestive failure). So, I took her to surgery and quickly put her on bypass. When I opened her left atrium I found that her cusps were soft and pliable, her valve had a pinpoint opening and I opted to separate the cusps by sharp dissection under direct vision, rather than replacing the valve with a mechanical one. At the end of this procedure, the valve looked competent; this is the advantage of operating under direct vision versus operating blindly with finger fracture.

The operation was without complication and Dr. Slotky, the cardiologist, took care of her discharge medication. On the tenth day, she was sent back to her family physician, but visited me monthly, and over the years we bonded. She became a friend to me and my wife.

THE FOLLOW UP

When her family doctor died five years later (he was a chain smoker), she decided to designate me as her doctor. I told her to stop smoking and monitored her blood pressure, cholesterol, digitalization, etc. Ten years later, she started experiencing dizzy spells. She consulted with many neurologists and ENT specialists, and some diagnosed her with inner ear disease. Others said it was her nerves, while still others claimed it was hyperventilation.

When she came back to me, I sat with her, got a complete medical history, and discovered she was still secretly smoking. Her pulmonary function test revealed COPD (chronic obstructive pulmonary disease). I suggested she have oxygen at home for one hour before she went to sleep and that cured her dizzy spells.

It is my experience that when you devote time for a detailed history, you solve the riddle and can give the right treatment. Taking the time to listen carefully to patients makes a big difference.

In 1995 she was sixty-five-years-old and started having tremors. She called me complaining she could not pour a cup of tea due to the severe shaking of her hands. She was diagnosed with Parkinsonism and various other neurological conditions, but on examination, her gait was normal and had no spasticity. She had a beautiful smile (not the facial mask of Parkinsonism). I had a strong feeling hers was a "senile tremor." I had previously read that beta blockers could reverse senile tremors and so I prescribed 25mg of Atenolol daily for her. This quelled her shakes completely.

In the last three years, she started suffering from heart failure. The usual treatment with Digoxin, Lasix, after-load reduction (recommended by her cardiologist) etc. could not control her symptoms and he suggested to me to replace her valve. She was now 77 with COPD, and I wasn't sure how she would tolerate the surgery. So, before subjecting her to it, I wanted to try an alternative heart medicine. That time, I heard in a lecture by a Dr. Sinatra (Fellow of the American College of Cardiology), that he had experience with using CO enzyme Q10 in patients with terminal heart failure in the ICU with success. He explained that CO enzyme Q10 plus Creatine plus Magnesium improved the metabolism of the heart muscle and reverse intractable heart failure. Using this alternative method saved her from heart failure.

I think a doctor should use all his knowledge, whether it is traditional or alternative medicine, if there is proof its use will benefit his patient.

Last year, three years after I retired, I called to inquire about her health. She was still alive and content thirty-five years after the operation. She appreciated my calling even after retiring in 2006. The bond with my patients is so strong that many of them call me from time to time to ask about my health.

A SERIOUS MISDIAGNOSIS: CELINA'S STORY

A forty-nine-year-old Mexican woman was referred to me by a cardiologist for mitral valve replacement. She was married and her two daughters accompanied her. As she did not speak English, her daughters served as translators.

In 1979, I inserted a Bjork Shiley valve in her heart, and placed her on anticoagulants, as usual. On subsequent ultrasound tests, her valve lumen was found to gradually tighten, but enough to keep her symptom free. I was planning to replace her stenosed valve at a suitable time.

One evening, she suddenly started to gasp for breath. Her daughters took her to a nearby hospital whereby the ER doctor told the daughters she had suffered a "heart attack" and admitted her to the ICU unit with a diagnosis of an acute MI (myocardial infarction). The doctor never called to consult me. The next morning, her daughters appeared in my office to tell me what had happened. I immediately knew her valve had clotted, and alerted my team to prepare for open heart surgery for mitral valve replacement.

It was not enough that they did a fatal mistake in the diagnosis; they did not want to release her. They'd rather leave her to die than admit their mistake, claiming she was not in fit condition for transportation. It was hard to convince the daughters to sign their mother out against medical advice, and I took the risk of committing myself to that advice for fear she might die in the ambulance.

She arrived at my hospital in a near death condition, and was taken straight to the operating room. With little preparation, she was put to sleep, and I put her on bypass.

She started to stabilize gradually and I did not lose any time in opening her left atrium and replacing her clotted valve with the new "St. Jude's" valve. We slowly weaned her from the bypass when her heart started to beat on its own. The excitement started to wear off and I came out of the operation satisfied, but mentally, physically, and emotionally drained. This is the high price surgeons pay to save a patient when every minute can make the difference between life and death.

NOTE

A wrong diagnosis can lead to disaster. One of the biggest problems in the practice of medicine in the USA is lack of communication and this is a good example. Although the family told the ER doctor that I was her surgeon, he never contacted me to at least clarify her past medical history, and admitted her with a wrong diagnosis that could have resulted in her demise. If a patient goes to the emergency room of a hospital close to his home for something minor, they steal him (take him over) and perform on him, unneeded tests or even unneeded surgery with disastrous results.

One of my patients went to a nearby hospital for an asthma attack and again, without contacting me, convinced her that she needed an "emergency" open heart surgery. She died during that unneeded operation. Her son-in-law was my janitor at that time, and his wife (the daughter of the victim) suffered psychologically from the unexpected death of her mother. This happens so often and is always backed by poor excuses.

If a family doctor calls more than one specialist on a consultation, they rarely get in touch with each other or with him. They scribble some unreadable note on the chart and sometimes double order medications, or give medications which counteract one another. That is why I always wrote orders in my chart that no medicine should be administered by any consultant without my prior approval.

Not only should doctor-patient relationships improve, but doctor to doctor and hospital to doctor relationships as well. Doctors should learn to communicate better with each other and find the time to convene a conference and discuss a patient's condition to agree on a plan of action, especially if the patient has multiple system disease or suffers a serious complication.

This collective care will help protect against malpractice suits and result in the best possible treatment for the patient, yet it is rarely done in American hospitals, and only after the patient dies. Once a month they conduct the so called morbidity and mortality meeting, but never bring all the cases to learn from them. They often bring those of the doctors whom they want to embarrass. This is hospital politics, and it's shameful! So far, this unethical problem has not been addressed.

This is not the only case, but you'll read later about a similar patient in coma with a similar reaction of her hospital. I hope these vivid examples will lead to action from the AMA.

LIVING BY NUCLEAR POWER

INTRODUCTION

Corrected Transposition of the Great Vessels is a rare condition. The incidence of it is 40 per 100,000. It comprises 0.5% of all congenital heart diseases with total reported cases in world literature standing at 1,000. I was privileged to see one of them in my lifetime and I am very satisfied with the way I handled it. My patient was fortunate he did not have any additional anomalies, as 65% of patients with more anomalies died within ten years. If not operated on, the median age for survival is forty years.

JESSIE'S STORY

Jessie V. was a thirty-three-year-old salesman who was referred to me because of a slow heart beat since childhood. He'd recently started feeling more tired than usual after minimal effort. He was a highly motivated man with a wife and two kids. In high school, he was told his heart was on the right side but nobody ever bothered to investigate or correct it. He had to work harder than other students to make his grades, wishing he had more energy and endurance. He was tall and thin with an appearance befitting to his age. His pulse was at 33 beats per minute, with blood pressure at 100/60. His temperature and gait were normal. His apex beat was best heard in the right sternal margin. The rest of the physical exam was normal.

His EKG was abnormal. His pulse was much too slow with multiple PAC's (premature atrial contractions) and some PVC's (premature ventricular contractions). A chest x-ray revealed a shift of his heart shadow to the right with a narrow outlet of the great vessels.

I admitted the patient to the cardiac unit in Mt. Sinai Hospital whereby the cardiologist, Dr. Slotky, performed a cardiac catheterization with me standing by for an emergency pacemaker in case the heart stopped. At that time there was no other diagnostic tool. The ultra sound, CT scans and MRI's were under development. The conclusion was that he had a corrected transposition of the great vessels causing disturbance in his conduction system that caused his slow heartbeat.

The patient needed a pacemaker. If left in this condition the heart rate decreases with age and ultimately stops, but a pacemaker allowed him to lead a normal life. The question now was what type of pacemaker.

Of the different types of pacemakers available I felt that the transvenous was contraindicated, so I opted for the epicardial. Due to the inversion of the ventricles I couldn't tell which side of the chest I should open to position the epicardial electrodes on the appropriate ventricle, so I opted to do it through a median sternotomy. The usual pacemakers at that time had a battery life of 5-6 years maximum. This meant that the chest has to be reopened for the pulse generator to be changed every five to six years during his lifespan.

Because of my interest in the advances in pacemakers and their electrodes, I came to know that Medtronic had just developed a nuclear pacemaker, which would last as long as the patient lived and did not need to be replaced at any point. Given Jessie's young age, I ordered the nuclear pacemaker.

Through the medial sternotomy, I could easily identify the right ventricle where the muscle is thicker and more suitable for implantation. To facilitate implantation, Medtronic invented screw-in electrodes which could be screwed into the ventricular muscles without sutures, as in the old models.

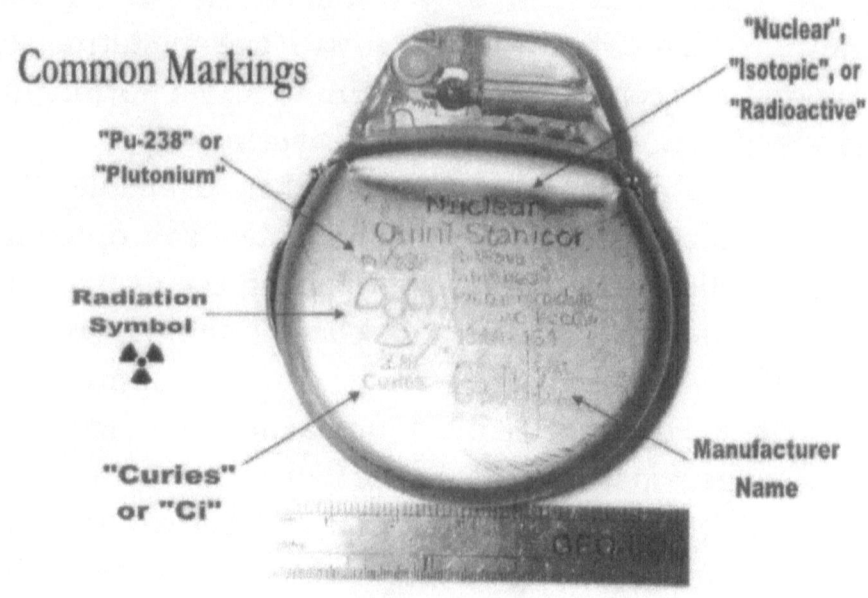

Common Markings

"Nuclear", "Isotopic", or "Radioactive"

"Pu-238" or "Plutonium"

Radiation Symbol

"Curies" or "Ci"

Manufacturer Name

Figure 20

Pacemakers are used to stimulate a regular heartbeat when the body's natural electrical pacing system is irregular or not transmitting properly.

Over the years, various power sources have been used for pacemakers, including thermoelectric batteries containing 2 to 4 curies of plutonium-238 (88 year half-life). As the term *thermoelectric* implies, the heat from the decaying plutonium is used to generate the electricity that stimulates the heart. As of 2003, there were between 50 and 100 people in the US who have nuclear powered pacemakers. When one of these individuals dies, the pacemaker is supposed to be removed and shipped to Los Alamos where the plutonium will be removed.

The Medtronic pacemaker shown in Figure 20 has had its plutonium removed.

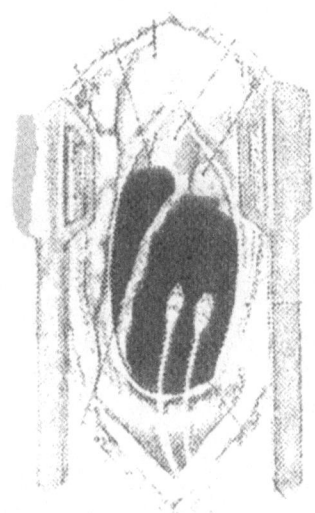

Figure 21: Implantation of Epicardial Electrodes

Two electrodes were screwed into the right ventricle and a bipolar nuclear pacemaker was connected to them and adjusted to a pulse rate of 72 beats per minute. This was then placed under the pectoralis muscle and the sternum was closed. The operation ended successfully.

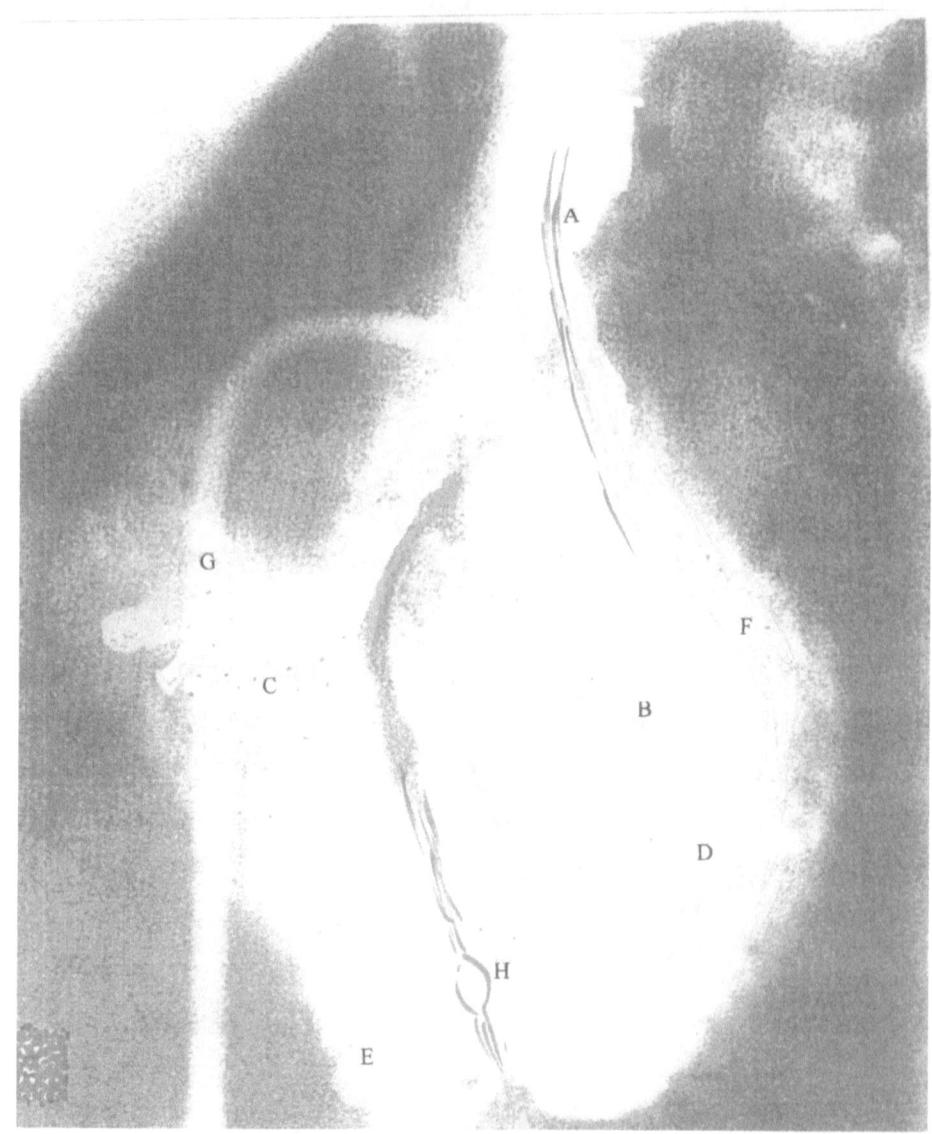

Figure 22: Diagram of heart.
A: Aorta. B: Right ventricle. C: Left ventricle.
D: High Pressure. E: Low pressure. F: Thickened ventricle wall.
G: Pulmonary artery. H: Ventral septal defect.

As you see in Figure 22, in corrected transposition, the right ventricle is in the left side, the aorta comes out of it and it assumes the function of the left ventricle (stronger). Hence its muscle is thicker. This is why I inserted the electrodes in it. I was glad that I used a median sternotomy in this complex situation.

Post-operatively I saw the patient a few times, and then he suddenly disappeared. To my great surprise he did not come to my office again for a follow up. Later, I learned that, without informing me, the Atomic Energy Commission decided to put him under the supervision of Dr. T. Baffas, the chairman of the Department of Cardiovascular Surgery.

ANOTHER BIG SURPRISE

Wonders never cease. In 2003, thirty years after I implanted his atomic pacemaker, he appeared in my office. By this time, Dr. Baffas had died and the Atomic Energy Commission wanted him to continue under supervision, so he made an appointment with me. I did not recognize him until he introduced himself because he had put on some weight. He told me that he had divorced his first wife, remarried, and had been promoted to manager.

His pulse was still 72 beats per minute and his blood pressure was 130/90. He was so grateful because nobody offered him any help during the first thirty-three years of his life.

"I was born again with new energy and zeal after you inserted the pacemaker," he said.

I humorously replied, "You became a bionic man."

Because of the complexity of his disease, nobody offered him help. It was obvious that the Trans venous pacemaker, which was the method common at that time, was not a safe option for him so the epicardial atomic was the one of choice. The epicardial approach went out of fashion at that time and it was his good luck that I was trained in the era in which it was commonly used, and I could use my experience from that time when needed. To my knowledge, Jessie V's was the only atomic pacemaker inserted in Chicago to date, and the fourth in the USA, possibly worldwide at that time.

NOTE

He's now 63 and still alive and well without any further interventions, thanks to the atomic pacemaker and my daring choice to do it because of his young age.

Epicardial pacemakers were usually inserted through a left thoracotomy. Median sternotomy isn't the usual method, but in this case it was the best choice because of his heart's complex anatomy. We don't have to stick to the "usual and customary," but deviate from it if there is a better way.

In 2006, I retired and referred him back to the chairman of surgery at Mt. Sinai Hospital. I was very gratified to see him for follow up before I retired though, and that I could make a difference in his life.

As for me, I am grateful that during my medical career, I was privileged to see and treat two very, very rare diseases: the myoblastoma of the lung and the corrected transposition of the great vessels.

SECTION IV
HIGH RISK PATIENTS DENIED PROPER CARE

HIGH RISK PATIENTS

"It is our job to give the best treatment possible. It is God's job to execute it."

—Charles Baker

INTRODUCTION

High risk surgical patients are denied surgery because of the fear of litigation if the surgery is not successful. This completely ignores the benefit of the surgery if it succeeds. Without risk there would be no success. Hospitals often block the surgery that could be the patient's only hope for survival. That is only one of the calamities that result from the epidemic of malpractice on healthcare. It raises cost, decreases efficiency, and results in patients dying without surgery.

THE MEXICAN COWBOY: JOSE'S STORY

Jose was a fifty-five-year-old Mexican farm owner who suffered from emphysema and high blood pressure, which were controlled by medications. He was tall and slim, dressed like a cowboy, with long boots and a hat. His passions included horseback riding, tractor driving, farming, and shepherding. He was a very pleasant and grateful man with three devoted sons who alternated bringing him to my office whenever he found himself in respiratory distress.

His treatment started with cessation of smoking, chest PT, nebulizer therapy, bronchodilators, aspiration of mucus, oxygen when needed, and calcium channel blockers to control his hypertension.

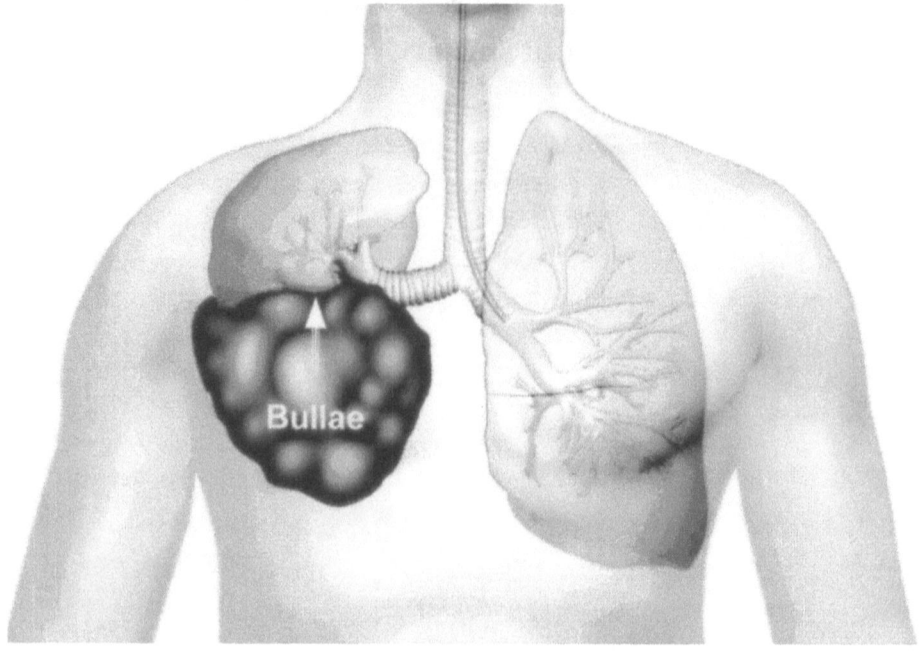

Figure 23

One day, his sons brought him to my office in a wheelchair with an unusual attack of difficult breathing. He was breathing rapidly and was somewhat cyanotic. His arterial blood gases revealed low O_2 and some CO_2 retention. An emergency chest x-ray revealed the cause of this dramatic change: one of his blebs grew into a big cyst in the right lung compressing it to a small volume with little or no vital capacity, and his left lung was severely impaired with COPD. There was not enough oxygen to meet his needs, so he hyperventilated to compensate. His EKG revealed sinus tachycardia.

After emergency lung care, he was transferred to Thorek Hospital by ambulance and I tried to schedule him for an emergency surgery to excise the cyst in order to decompress the right lung. In the meantime, he was placed on oxygen by cannula through his nose. With this, his color improved and his pulse became regular.

Fearful of being blamed or sued if something happened during surgery, the anesthesiologist reported the gravity of his condition to the hospital's medical director who then obstructed the surgery. Although he was placed on oxygen, he was unable to get out of bed and walk to the bathroom. His health was severely impaired, his breathing labored, and the only chance for his survival was to excise the cyst and let the compressed right lung rise up to fill his right thorax.

No matter how much I explained his incapacitating difficulty to breathe to the medical director and the anesthesiologist, they were adamant in their refusals to give permission for surgery; they would rather have the patient die than take the risk.

Miracles always happen. Brokenhearted to leave him alone under the care of a physician in our hospital, I accepted the invitation of my dear friend and colleague, Dr. Bangash of New Jersey, to attend his yearly Christmas party. By mere chance, I was seated next to a chest physician from my friend's hospital and related to him the story of my patient, the Mexican Cowboy. This very astute chest physician told me to tell my medical director that if the cyst ruptured and the patient died from a tension pneumothorax, the hospital would face a bigger lawsuit for not allowing me to operate on him. Ultimately, the risk of doing nothing is greater than the risk of operating. He gave me his telephone number and offered to talk to the medical director if needed. Hearing that, I immediately flew back to Chicago and relayed this message to the medical director. At first, he did not believe me and asked for the doctor's telephone number.

After speaking with the chest doctor from New Jersey, the medical director gave permission to proceed with surgery—not for the benefit of the patient, but to avoid a possible bigger malpractice lawsuit.

The anesthetist, under the cover of "informed consent" tried to scare the patient into not accepting surgery. He told him that he might die in surgery or remain on the respirator forever, or leave surgery in worse condition.

But no matter what the anesthesiologist told him, the patient's response was always, "I trust my doctor." He signed the consent without hesitation.

THE SURGERY

The surgery went surprisingly smooth. I had the anesthesiologist give anesthesia only to the left lung because adding more air to the right lung might cause the cyst to rupture or increase in size. Rather than excising the cyst, which would leave a raw surface on the expanded lung, the idea came to me to plicate the cyst into a thin covering over the expanding lung; surgery took less than 20 minutes.

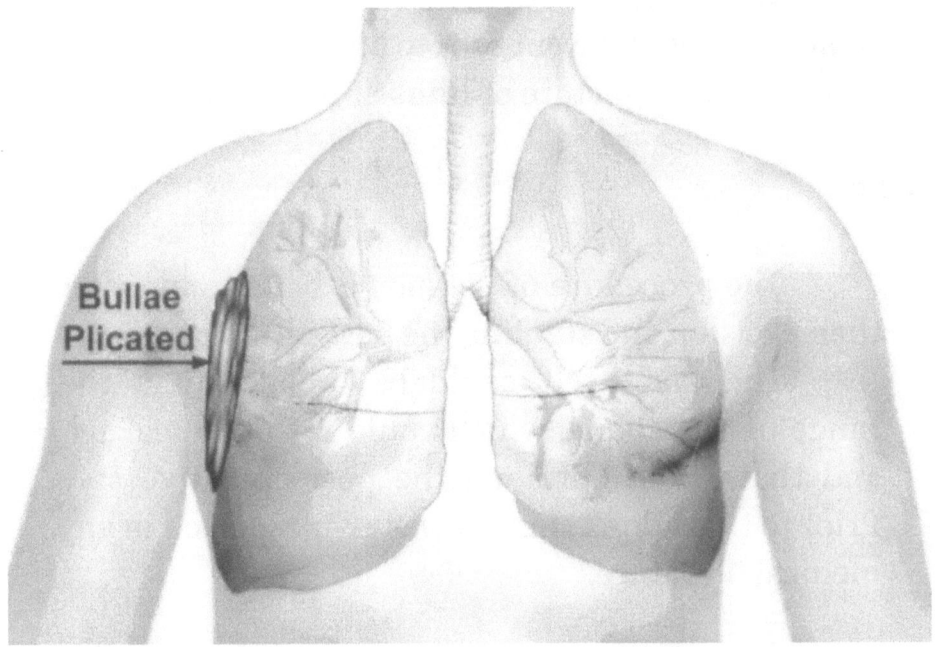

Figure 24: Plication of the cyst is better than stapling and excision of the cyst. There is no leakage between the staples.

THE POST-OPERATIVE COURSE

With the right lung expanding with no air leaks, the chest tube was removed the next morning. The patient was gradually weaned off the respirator, and a day later he was extubated and could breathe on his own. A week after surgery the stitches were removed and Jose was discharged with improved breathing.

Armed with the success of the operation, I told the medical director, the administrator, and the anesthesiologist, "You were going to sacrifice the life of the patient for what you call protecting your hospital. You should always think about protecting the patient more."

A month later Jose went back to Mexico. It was gratifying when he sent me a picture of himself driving his tractor.

NOTE

The procedure of *plication* is better and faster than excision and stapling the base of the cyst.

The cyst itself could be broader than the range of the staple, which could result in leakage, and that would prolong the post-operative stay. Prior to this I don't think it was ever reported in any medical literature.

HIGH RISK PATIENTS AND MALPRACTICE

INTRODUCTION

High risk patients are being denied surgery because of fear of malpractice. This is one of the ways that greedy trial lawyers are impairing medical practice and obstructing the care of people who might otherwise be saved.

Medical patients as well, especially those with multi-system diseases are denied proper medical care. They cannot find a family doctor to take responsibility for their care because the general practitioner is so paralyzed by the fear of litigation that, rather than taking care of them himself, he refers them to several specialists, one for every system affected. He doesn't do this because of lack of knowledge, but rather because he wants to cover himself if he is sued. This results in severe discomfort to those unfortunate patients who don't have the means of transportation to visit so many doctors. It can create a tremendous burden on the caretakers who usually accompany them during those visits.

The malpractice epidemic is causing a severe rise in the cost of medical care. It also decreases efficiency and increases the mortality to both medical and surgical patients who could be saved by proper and effective means. The reason for that is the trial lawyers have a strong lobby and, from the lucrative rewards they get from litigation, they can afford to make substantial donations to some lawmakers. In return, those lawmakers block any motion for torte reform.

It is amazing to me that all discussions of healthcare reform ignore the epidemic of malpractice in the United States more than any other industrialized country. Some doctors do not function to the best of their abilities. Gynecologists refuse to deliver babies, and in some rural areas, a pregnant woman may have to travel 100 miles to find a doctor to schedule her delivery. Medical care is a basic human need.

MULTI-SYSTEM DISEASE PATIENTS ARE CONSIDERED HIGH RISK—REFUSED FOR PROPER CARE: JANET'S STORY

Janet was fifty-years-old when I performed a huge hiatal hernia repair on her. The operation was risky and hazardous because she suffered from COPD and was overweight. Post-operatively, she needed physical therapy and rehab to get her out of bed, walk her, and address her exacerbation of COPD. The good thing on her side was that she was not a smoker herself. In addition, she had type II diabetes (adult onset, as opposed to type I, or juvenile diabetes, where the patient is born with it and the only treatment is insulin) and an enlarged heart. In spite of all this, she had a very positive attitude.

Because Janet had inaccessible veins, I inserted a porto-cath for her and used it to medicate her with Aminophylline, and sometimes Prednisone whenever she had an asthma attack, or exacerbation of COPD, supplementing that with some nebulizer therapy. She would leave my office breathing comfortably.

As for diabetes, I rarely give insulin to type II diabetes (late onset); they're insulin resistant anyway. I recommend oral Glucovance before each meal and Actos at lunchtime. Surprisingly, this regulates their diabetes, as proven by daily home accuchecks.

As far as Janet's arthritis was concerned, I learned about new intra-articular injections of Hyalgan or Synvisc, which proved to be of amazing benefit and on many occasions, spared patients surgical knee replacements. These injections do not need an orthopedic surgeon to administer them. All it takes is to feel the space between the tibia and femur on either side of the patella, and inject in that space. I gave Janet five such intra-articular injections in every knee, a week apart, and a miraculous improvement happened. The patient was able to give up her wheelchair and walk. Sometimes she walked with a cane, other times with a walker or the support of her daughter. I read recently that the five injections have been concentrated in one long acting injection and I think it should be tried before considering knee replacement.

The technique of intra-articular injections is easy and could be taught to any physician or nurse practitioner. Many elderly patients who develop arthritis in the knees improve tremendously with them.

As for Janet's cardiac dilatation and heat failure, she responded well to the usual treatment with Lanoxin, Lasix and Vasotec as an after-load reducer. Added to that, I gave her Co-Enzyme Q-10 to enhance the metabolism of her heart muscle (The Sinatra Technique).

NOTE

A month after my retirement, she called me in desperation with an attack of COPD. I advised her to be admitted through the ER under a family physician. This physician called five specialists to take care of her COPD, arthritis, diabetes, heart failure and obesity. Upon her discharge, he advised her to leave her home, where she was helped by her daughter and neighbors, and admit herself to a nursing home. She was told, "You have one leg here and one leg there," meaning that she was between life and death. This made her very depressed.

I believe such destructive comments are out of place in this age of mind-body medicine.

I added her story to bring attention to the problem of patients with multiple-system diseases. They cannot find a family doctor to accept the responsibility for their care and refer them to multiple unneeded specialists for fear of litigation.

Maybe if we expand the medical education to six years instead of four, as in Europe, graduates would have the knowledge and confidence to treat such patients and worry less about the possibility of malpractice suits.

SECTION V

PATIENTS IN COMA

PATIENTS IN COMA

INTRODUCTION

Patients in a coma are closest to my heart because they need very special care since they cannot speak for themselves. Their lives depend on the care givers. The physician needs to be vigilant in reminding the staff that these patients are viable and nursing care is the most important aspect. They require extra attention and time compared to the usual patients. Special hygiene, such as shaving the patient to make him look good for his visitors, and the usual turning from side to side to avoid bedsores, are all required tasks from the nursing staff. Frequent suctioning if the patient has a tracheotomy, together with watching and adjusting the respirator are all factors that play a role in his survival. Their lives are in their hands. The doctor's enthusiasm, love and attention to every detail inspire the staff to do likewise.

The most important component to keeping these patients alive is nutrition, because they cannot eat or drink by themselves. The patient is usually put on a liquid diet delivered through a gastric or intestinal tube. Some of these supplements, like Ensure, can cause diarrhea (Liquid in, liquid out).

The blood's oxygen level should be monitored daily and the respirator adjusted to avoid brain anoxia (lack of oxygen in the brain). Tabulating the blood test results and patient's weight, along with the intake and output, facilitates the

151

daily comparisons with previous days, so that from one glance at the chart, the doctor can know if the patient is progressing or regressing and if so, what could be done.

GIVEN THREE DAYS TO LIVE: ADELPHA'S STORY

Mr. Camacho, a policeman, and all his family were my patients. As a policeman, Mr. Camacho was honored for his diligence, kindness and courage, and he won my respect. His squad car partner was Mr. Silvas. That afternoon, Mr. Silvas was preoccupied, sad and distracted. This worried Mr. Camacho who asked what was bothering him. Mr. Silvas, with tears in his eyes, told his partner the sad story of his wife always being subject to pulmonary embolisms. At present, she was hospitalized in a facility on the south side of Chicago. In spite of all the treatments they gave her, she went into coma and had been on a respirator for the past three weeks. That day they told him that there was no hope for her recovery and no medical options left, and that they were arranging to send her to a Vencor hospital where patients remain on artificial respiration until they die.

It was night when, upon hearing that sad story, Mr. Camacho suggested to him, "Why not call my doctor for a second opinion?" Mr. Camacho called me at about 8:00pm as I was about to leave my office and asked me to speak with his partner.

Mr. Silvas, spoke to me while holding back his tears and related to me, the story of his wife who was given only three days to live. I was hesitant to give any opinion about a patient who I had never seen nor examined before, but Mr. Camacho asked if I could have her transferred to my hospital instead of a Vencor hospital. I hesitatingly agreed to it. I don't know why; maybe not to disappoint Camacho, maybe for a feeling in my heart that *maybe something more could be done.*

This created a lot of turmoil. Her hospital supervisor said she had received the maximum care and there was nothing more to do for her. She also told my hospital's admitting office that the insurance might not pay for another hospitalization for the same sickness, which made my hospital reluctant to take her.

Compassion and empathy told me I could do something different to save her life, especially when her husband told me she was the mother of five teenage girls, and they'd all be lost if she died. With that, he won my sympathy. An inner voice pushed me to find something to do to help her.

Luckily, the nursing supervisor that evening, who was assigned to screen the admissions in my hospital, had faith in me, so she accepted to admit Mrs. Silvas through the Emergency Room when she arrived by ambulance. I asked the ER doctor to do the usual chest X-ray, CBC, coagulation profile, emergency ventilation perfusion lung scan, and arterial blood gases while I was on my way to the hospital.

When I arrived at the hospital and went to see her, she was in a deep coma and unresponsive to any stimulation. I reviewed the blood gas results, which revealed poor PO_2 and high PCO_2. This indicated to me she might be in CO_2 narcosis and that the respirator was not well adjusted. The lung scan revealed scattered emboli all over both lungs. I admitted her to the ICU and adjusted her respirator to give her more oxygen and increase the frequency of ventilation to wash out her CO_2. I thought all night as to what more I could do for her. I considered inserting an IVC (inferior vena cava) filter but I learned that she had one in the previous hospital which did not prevent further embolism, so that was not an option. I had to think beyond that.

In the morning, as soon as I arrived at the hospital, the receptionist told me that the medical director and the hospital administrator wanted to see me before I began my rounds. As soon as I entered the medical director's office, I was met with gloomy faces.

The medical director started by saying, "What is the reason you admitted a dying patient in this hospital? You know that we are neither Vencor nor hospice. She received maximum care at the other hospital. Did you bring her to die here under your watch?'

"No Dr. T," I said, "I brought her here because I had a feeling that I could do something more for her."

Then the administrator jumped on the wagon of accusations. "If you have such sympathy for her and wanted something more to be done, why didn't you send her to a university hospital?" he asked.

I looked at him and said, "This is demeaning to this hospital to think that a university hospital could do better than us. The family put their trust in me and not in the university. I know I am putting my reputation and credibility in jeopardy here, but if I could save her life, that would be worth my while."

The medical director held back, impressed with my attitude. "Josh, go see your patient and keep me informed."

Saddened by the whole situation, I took her chart, went to a consultation room and closed the door. Praying for inspiration, I asked myself, what more can I do for her?

In what now seems like a trance, a statement which I'd heard fifteen years back at a conference, jumped into my mind. Dr. William Blaisdell, a famous vascular surgeon from

San Francisco, observed that our coagulation profiles were not accurate and that the amount of Heparin we gave in those situations were only prophylactic, but not enough to dissolve an embolus or a thrombus and that to do so, we must almost "double" the amount of Heparin. He tried this on fifteen brachial embolisms that came to the ER of his hospital in San Francisco and were all cured without surgery or complications.

With this in mind, I went back to the patient and corrected a leak around her tracheotomy by purse string suture and then ordered, in addition to what was given in the previous hospital, 4000u of Heparin STAT and 1000u more every hour.

Our oncologist/hematologist, Dr. Chawla, was convinced I'd lost my mind. I told him where to find the article, and convinced him there was nothing to lose, and that according to the other hospital report she was already dying. He reluctantly supported me.

The next morning, the patient miraculously opened her eyes and looked around, then closed her eyes again and slipped back into a coma. Despite that, this was a very encouraging sign. I was on alert over the next twenty-four hours for any bleeding complications from the high heparin dose which luckily never occurred.

On the third-day post admission, she opened her eyes for a longer period of time and said, "Where am I?" Am I still alive?" and closed her eyes again. This time, her blood gases were much improved.

It was encouraging for me when I saw her swallowing her saliva so I asked the nurses to try to give her some coffee with a spoon. They hesitated; afraid she might choke, so I did it myself. This encouraged the nurses to continue to give her sips of water and coffee for the next twenty-four hours. To be sure of the improvement, I ordered another ventilation perfusion lung scan which showed great improvement in her pulmonary vasculature, meaning less emboli. In addition I continued to administer high-dose heparin.

The following day, she seemed more alert when I came to see her. She asked who I was and then said,

"Where am I?" I told her she was in a different hospital. She closed her eyes for a moment, opened them again, and asked, "Am I still alive?"

When I told her it was the fourth day, she asked, "How long have I been here? I heard with my own ears that I had only three days to live. I was even crying that I wouldn't see my daughters' weddings so, is this a temporary awakening or am I still going to die?"

I assured her that we were doing everything to keep her alive and that so far there was a great improvement in her lungs. And as I gave her more sips of coffee, she said, "Oh how nice. I hadn't tasted coffee for three weeks."

At this point, I put her on assisted ventilation, which means the machine will kick in if she stops breathing or cannot breathe enough on her own. On the fifth day, I asked her if she wanted to get out or bed. After glaring at me for a while she said, "Can I?"

"Yes, if you want," I said.

When I asked the nurses to get her out of bed, nobody believed she could sit on a chair. They all said that they were afraid she might pass out or fall down. So I gave the patient a "bear hug" and practically carried her to a chair next to the bed and took her off the respirator, but left her breathing moisturized oxygen.

On the sixth day, her blood gases were so good without the respirator that I removed the tracheotomy tube and told her she was alive and breathing on her own. I ordered physical therapy to begin rehabilitation, so they stood her up and made her take a couple of steps around the bed. The process of walking her was gradual and at the end of the next four days, I asked for a new ventilation perfusion lung scan which revealed her lungs were clear.

When I called her husband and told him to prepare to take her home, he did not believe me. I told him to thank Mr. Camacho for referring her to me and to thank God for her salvation.

At this stage, we started weaning her off the heparin and adjusting the dose of oral Coumadin.

When it came time for final discharge, I asked the medical director to co-sign my discharge order. He was, of course, happy as this was good for the reputation of his hospital, but teasingly he said to me, "Josh, you were lucky."

I don't think it was a matter of luck, but rather my compassion and determination not to disappoint someone who had put faith in me. Constantly looking for advances in our field, and staying abreast of medical literature, is important for success. What you hear at a conference today might help at a later date.

Her husband and daughters were around her and Mr. Camacho called to thank me. I told him to thank God for inspiring me to do what I did.

When she came to see me in the office, my nurses who knew the story were happy to see her. That year, my nurses threw a birthday party for me in the atrium of my office building and invited her. There were tears in my eyes when I saw the woman who had once been condemned to death dance Merengue for three hours almost nonstop. That was the highlight of my birthday.

Whenever you are faced with the response, "We did all we could and there is nothing we can do to save a life," ask yourself, *"What more can I do?"*

NOTE

This is another example of the reluctance of hospitals to accept high risk patients out of fear of malpractice. There are difficulties with insurance companies, especially when a patient is transferred from hospital to hospital for the same condition; the insurance company must be convinced the patient will receive more than what they did in the first hospital. In Mrs. Silva's case, they had to pay in full, because that first hospital condemned her to death and our hospital brought her back to life.

Sometimes, a doctor has to put his reputation at stake if he thinks he can save the life of a patient. Faith, trust in God, and the courage to accept blame when something goes wrong, are needed to achieve miraculous cures. Armed with these qualities, a surgeon should be willing to accept reasonable risks if he feels in his heart he can save the patient, because without risks, there are no successes.

REFERENCES:

William Blaisdell et al "Management of Acute Extremity Embolism and Thrombosis, published in *Surgery 1978* 84.822-834.

William Blaisdell et al "In vivo Assessment of Anticoagulation" published in *Surgery 1977* 82.827-839

Current Therapy in Vascular Surgery, 2nd Edition by Ernst-Stanley page 939, "In life threatening conditions such as pulmonary embolism a high dose of 20,000u bolus followed with 4,000u of heparin/hr. when the symptoms subside, cut back over 3-4 days to modest levels."

SAVED BY STARBUCKS: PHIL'S STORY

Phil was a thirty-eight-year-old chronic alcoholic who defied all efforts to stop his drinking. He was referred to me by his father on whom I had performed a coronary bypass years prior. The father was a recovering alcoholic himself and never touched a drink or accepted a sedative after surgery. Phil worked as a dishwasher in a nursing home and was loved by his coworkers. He was admitted to the hospital because of massive hematuria. Several urologists failed to find the cause. The cystoscopies revealed only blood, although his coagulation profile was normal. He received massive blood transfusions to replace his loss. At this stage, however, he was still able to eat a normal diet.

He was also suffering from alcoholic cardio myopathy, and in spite of balancing his intake and output with a swan-ganz catheter he went into heart failure and was unable to breathe. He needed to be intubated and put on a respirator.

To add to the poor man's problems, he acquired pneumonia and we had to give him massive doses of intravenous antibiotics. In spite of these treatments, he slipped into a coma. All the consultants which I recruited to help me in his care gave up and said he was hopeless and there was no way to save him. I thought differently.

He was young and it would be a blow to his father if he died. It was difficult to motivate the nursing staff not to give up on him and treat him as anything other than a dying, hopeless person.

A tracheotomy was needed to replace the endotracheal tube. I succeeded in relieving him of pulmonary edema

161

and cured his respiratory infection. He was no longer bleeding from his bladder, but then came the question of feeding him.

An abdominal surgeon was called by me to insert a jejunostomy feeding tube and put him on the nutritional feeding protocol of the hospital: baby food every four hours. However, under this regimen the patient was losing weight and it wasn't from fluid loss. After ten days in coma and on this nutritional program, he was practically skin and bones. I was afraid, after all the progress, we would lose him to malnutrition. He was wasting away.

Leaving his room sad and desperate, I went to the cafeteria, though I had no appetite to eat. I opened the refrigerator and grabbed a bottle of Starbucks cappuccino.

After sipping it little by little, I turned the bottle around and realized it contained 220 calories. I thought it was fattening for me but perfect for a guy like Phil. I figured if we fed him this instead of the baby food, it would give him a little more than 1200 calories a day, which he needed for his bodily functions; the caffeine might also help wake him up. I ran upstairs and wrote an order to give him Starbuck's cappuccino every four hours instead of the hospital supplements.

Following that order, all hell broke loose. The dietician, unfamiliar with this new treatment, ran to her supervisor, the senior dietician, who was reluctant to follow up on my unusual order. She ran to the director of nursing who had also never heard of this before. She stormed into the administrator's office, told him, and he went to the medical director to cancel my order.

He claimed he had to protect the hospital from malpractice that could result from administrating something which was not "usual and customary."

My last resort was the chief of the medical department, the same Dr. Chawla who helped me with the heparin dose on Mrs. Silva. I convinced him there was nothing to lose and much to gain by what I suggested, and that watching a patient losing weight until he dies from malnutrition is more of a liability for the hospital and grounds for malpractice than my suggestion.

Dr. Chawla was luckily very open-minded and understood my point. He went to the medical director and told him there was nothing wrong with my order and that it was in the hospital's favor to give him the caloric count he needed, which hospital food was not providing.

My orders were finally approved. By now, he had been in a coma for three weeks. Three days after administering the Starbuck's cappuccino, the patient opened his eyes and looked around. The second day, he requested food and I started him on a liquid diet, which gradually progressed to a soft diet. I recommended he dangle his legs while in bed, and a day later, sat him in a chair. The physical therapy and the rehabilitation departments were then called to start him exercising, and gradually walking him around the bed, and then through the corridors. Ten days after starting the Starbucks diet, tube feeding was discontinued and the patient was discharged. He was breathing, eating and urinating normally, and it was all thanks to a hunch.

COMMENT

Phil went back to work in the nursing home, and his coworkers received him back with a party. His father was grateful for his salvation, but disappointed that, unlike him, Phil was secretly drinking again. He blamed his "friends" for tempting him. Every time he came to visit with his father, I counseled him.

I contacted Starbucks to publish his story. At first, they wanted to do it, but after some time, declined the idea. They thought that nobody would believe the story and people would think it was only propaganda!

SECTION VI

SERIOUS TRAUMA

CHILD PINNED UNDER A CAR: YVONNE'S STORY

I was coming out of Tel-Hashomer hospital at around five in the evening when I was shocked to find a Volkswagen on its side with a young girl, about ten-years-old, pinned underneath it. I got out of my car and still don't know how I got the strength to lift the Volkswagen from being rolled over on its side.

I pulled the child out from under and carried her in my arms to the backseat of my car and sped to the emergency room. It was the closest place equipped for trauma and much better than waiting for an ambulance to carry her.

Upon calling a "code blue," the emergency staff immediately came to her aid. A young anesthesia resident visiting from England was on call and immediately intubated her. We took a quick chest x-ray and blood for cross match and had the blood bank on alert for any amount of blood she needed because her hemoglobin was very low. It was clear the abdomen was distended and bleeding profusely. The operating room was alerted to prepare for surgery; and X-ray revealed fractured ribs on both sides; the lungs however, were expanded.

In the operating room, she was quickly prepped and draped and the abdomen was opened where she was bleeding profusely. Her blood pressure was as low as 80/60. The torn spleen, which was the main source of bleeding, was quickly removed and the crushed part of the left lobe of the liver was quickly excised, and the raw surface closed. With blood transfusion, and stopping these two sources of bleeding, the blood pressure went up to

100/60. The nonviable part of the intestine was resected and I started reconnecting her bowels.

THE BIG SURPRISE

While inspecting the abdomen for other sites of bleeding, the young anesthesiologist dramatically threw the drapes away from the patient's chest and shouted in panic, "She's dead! I can't hear any heart sounds! And look at the blood pressure! It's down and the heart is beating erratically on the monitor!"

Although shocked, I told myself to keep cool. It suddenly dawned on me that her broken ribs had pierced the right lung and with the anesthesiologist blowing air under pressure, she probably developed a tension pneumothorax that pushed the heart to the left, twisted the major blood vessels, causing this dramatic change.

While the young, inexperienced anesthesiologist was about to walk out, I shouted at him to stop and wait.

SAVED BY THE NEEDLE

I knew that if my hunch was true, a wide bore needle pushed into the pleural cavity would temporarily relieve the tension pneumothorax.

I asked the circulating nurse to hand me a wide bore needle. I inserted it in the right chest and ordered the anesthesiologist to continue bagging the patient and ordered an immediate chest x-ray. The x-ray confirmed my doubts; it showed the fractured ribs, retracted right lung, and the pneumothorax. I quickly inserted a chest tube and as usual put it under water drainage. We could see the air bubbles coming from the chest and gradually the monitors showed return to acceptable rhythm, and the anesthesiologist could hear the heartbeat again.

We re-draped the patient. The drama was over and I continued to explore the retro peritoneal hematomas and cauterized the ligaments of the spleen whereby the bleeding almost stopped by pressure. The liver stopped bleeding, and I persevered in cauterizing any oozing vessels and drained the retro peritoneal hematoma. The kidneys did not show signs of injury, and the patient had good urinary output. I closed the abdomen in layers without leaving a drain and admitted the patient to the ICU unit.

We left her on the respirator for a few days because her breathing was paradoxical. (The two lungs do not expand at the same time, but alternate individually because of the many rib fractures on both sides, making the chest wall unstable.)

Once stabilized, she was extubated, the chest tube was removed, and although most of the trauma was abdominal, I asked my boss, Dr. Pawzner, to admit her in our thoracic floor under my personal care. He was very compassionate and agreed. It was at this time that I got to meet the young girl's parents, both of whom were very grateful. They were not in the car that flipped. It seemed their daughter was crossing the street when the driver of a Volkswagen car made a sharp turn and flipped over landing on top of her. They knew she was almost dead when I snatched her from under the car. They felt I was in the right place at the right time. They knew that a few minutes of delay would have made the difference between their daughter's life and death.

This, however, was not the end of the poor girl's rehabilitation. She also had a fractured jaw, lost some of her teeth and had twisted and broken her nose. She was transferred to the maxillary/facial unit whereby she needed reconstructive surgery. Naturally, I visited with her daily and she was always happy to see me. I think it took 2-3 weeks more for her to recover from her facial injuries. It was a dramatic moment when she was discharged and I hugged her goodbye.

I heard the family lived in another city, so I naturally sent a discharge summary to their doctor. She visited with me a couple of times before I came to the United States to get more training in heart surgery.

Very often, I think of her, and now that I'm writing her story more than forty years later, I wonder if she went to college, if she has children, or if she even remembers what happened to her when she was young.

I thank God for the strength He gave me to lift up the car and for making me available to save her. The Lord wanted her to live and His blessing was in every part of her care.

NOTE

I discovered later that this anesthesiologist was a resident who was in Israel as an exchange student. It is an advantage for a surgeon to work with an anesthesiologist known to him and who is experienced and supportive during crucial times. Had I known he was only a resident, I would have asked for his senior to attend. In retrospect, I think he should've called for additional help.

When I have been faced with cases of patients near death, the image of what happened never leaves my mind. From time to time, I ask myself: What made me leave the hospital early at that particular time? Where did I get the strength to lift the car?

Questions of destiny and coincidence follow me to this day, especially when such experiences happened repeatedly. A research on this subject can be found in Part II.

"HERE'S MY ARM. CAN YOU PUT IT BACK?"

INTRODUCTION

Reattachment of severed limbs is one of the recent advances in surgery. Until World War II amputated limbs were replaced by prosthesis, which is rarely used now, as in cases where the amputated part could not be found or decayed from lack of oxygenated blood and became unsuitable.

The first successful re-implantation of an arm in the United States of America was in 1962 when a 12 year old boy was trying to jump onto a moving train in Boston and he was brought with his amputated arm to Massachusetts General Hospital. The senior resident in the emergency accumulated a group of surgeons: orthopedic, vascular and neurology doctors. The vascular surgeon anastomosed the vessels, the orthopedic surgeon fixed the bones and the neurology surgeon advised waiting for a second stage to attach the nerves. When that boy healed and started using his hand, he made big news and became a celebrity.

This opened the door to other centers in the USA and abroad to re-attaching severed limbs. In the years to follow, everyone at that time made news and was written up in the paper and shown on TV.

These operations became frequent and of great importance in the recent wars in Afghanistan and places occupied by terrorists whereby landmines and explosions blew away the limbs of our soldiers sometimes the two upper, others the lower and even all four limbs if the explosions were severe.

Fixing the two upper arms were of more importance for the future life of the patients every day functions, then the legs which could be replaced by prosthesis and needed for mobilization. Scooter and electric wheelchair industries helped for the purpose to move around the house and are taken even on cruises and other travels.

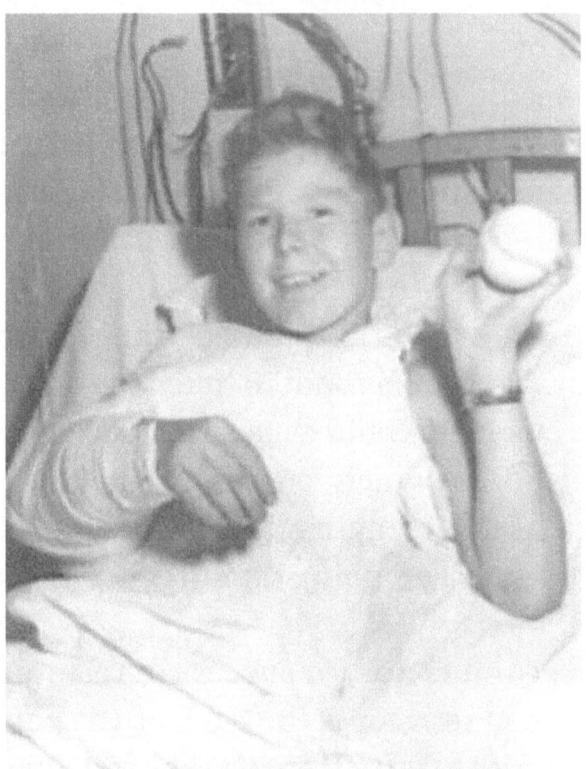

Figure 25

The biggest series of attaching arms, forearms, hands and fingers came from China's Shanghai. It covered their experience in 438 patients from 1963-1976 and described in detail the factors needed for successful reattachment:

1. The most important of which is establishing a blood supply to the excised part before it deteriorates and goes into rigor mortis. TIME between the injury and bringing the blood supply to the amputated part to keep it alive is of utmost importance.

In one recent case where the artery of the amputated arm was out of reach, it was reported that the paramedics inserted a needle in a leg artery and connected it with a tube to the artery of the amputated limb till they reached the center where anastomosis between the patient's artery and the severed artery was possible. If this happens in a remote place where the operative team was out of reach they could preserve the amputated arm in a sack of water cooled to four degrees but not to touch the amputated arm with ice because this could cause ice burns to its skin. The longest period in which successful reattachment was achieved between bringing the amputated part to an attachment center and doing the surgery was 3.8 hours.

2. The second factor for success in resuming a functional recover is the AGE of the patient. The younger the patient the shorter the bones which allows proximity of the blood

vessels and most important, the nerves
which grow an inch a month, could join fast
by themselves without a second operation
and that is all what is needed to regain
function in a child.

In one case of attaching a severed hand of a young boy, the surgeon removed one row of the carpal bones to obtain more proximity. I followed this child through his post op care and was surprised that he had full function of his hand in one month and could leave the hospital and do any sport or activity which requires using the hand and fingers. The longer the bones, the longer the distance between the severed nerves and that's why the neuro-surgeons like to wait a few months before trying to repair them or using a nerve graft to allow them to grow and shortening the distance between them.

3. It is *teamwork* and very rarely could be done
   by a surgeon experienced in vascular and
   orthopedic surgery.

4. *Experience*, especially in using magnification
   to attach fingers where the arteries are tiny
   and require a surgeon trained in working
   under a microscope.

DARIUS' STORY

While passing through the emergency room of St. Mary of Nazareth Hospital, I saw a man with an amputated limb at the level of the right mid arm hanging by a little piece of skin and another man holding the severed part for him. The upper arm was tied by a handkerchief to stop the bleeding.

I asked the emergency room doctor regarding what he plans to do with him. He told me that he called the emergency room doctor on call who told him that he has no experience in that kind of trauma. Out of desperation he called the chief of the surgical department and he told him to send the patient to a University Hospital. It took more than two hours for the operator to transfer him from doctor to doctor in one university hospital and they told him that they do not have emergency room service. He called another and got the same response. "So how long has this man been here?" I asked. "About 3 ½ to 4 hours" he said. "If this man is not operated on immediately he is going to lose his arm which becomes unsuitable for re-implantation." The man looked at me with tears in his eyes and said, "I do not want to lose my arm, please help me." I asked him, "How did this happen?" He said that his wife closed the door on him. I knew from my experience in Mt. Sinai that they never tell the truth.

Encouraged by a case of treating such an injury to the arm, presented by Dr.'s Bangash and Zaorski in a meeting of the Denton A. Cooley Cardiovascular Surgical Society, and having my heart feeling sympathy for him, I

176

volunteered to operate on him. The ER doctor immediately called the OR and seemed relieved from the inability to do something for this man. (Many university hospitals do not have emergency or trauma units and do not teach trauma to their students.)

In the operating room, I ordered my assistant resident to wash, with soap and water, both sides of the arm and prep the right arm and left groin to take the short saphenous vein used for coronary bypass as a bridge between the two severed arteries. This was advised by Dr. Bangash in his presentation so as not to pull the two segments and make an anastomosis under tension which will narrow when the patient extends his arm.

I started the operation, under general anesthesia, by taking a long segment from the short saphenous vein thinking that I might need it to bridge both arteries and veins. I used part of it to anastomose it at the proximal and distal ends of the brachial artery. I gave the patient 50 units of Heparin to avoid thrombosis. I was so gratified by seeing the man's nails pink and feeling a radial pulse. The veins were wide and elastic. They approximated and anastomosed easily. Then with 3-0 chromic catgut on a large needle, I started to suture the groups of muscle together. When it came to fix the bones, I could have left them approximated with a right angle plaster, but I preferred to call the Orthopedic Surgeon who was my friend and knew that I had experience in orthopedic surgery. He told me to try to insert a intra-medullary rod and if they don't have it, he said to put a three inch plate between the two segments of the bone and fix it with a couple of nails. This was easy. The neurosurgeons advised to leave the nerves to grow as much as possible and then anastomose them after the wound healed. I was gratified by restoring a living arm although with no mobility. Then I sutured the skin with 3-0 Dacron and applied a loose bandage and then I put the arm in a 90 degree angle and supported it with a plaster slab and I and my resident felt a good arterial pulse. I told the anesthetist to wake him up and you can't see the euphoria when he felt both of his arms still attached, but I preceded him by telling him he would not be able to move it yet, but not to worry because the neurosurgeon will attach his nerves at a later time after his wound heals and that it will take time and effort till he will be able to use his arm.

The post-operative care was long and tedious. Ten days after the operation I took off the plaster. The wound was healed so I asked the resident to remove the sutures and asked the physical therapist to flex and extend his elbow, wrist and fingers passively to avoid contractures and stiffening of the joints.

The patient had no insurance of any kind, not even public aid, but I didn't care. The greatest reward of a surgeon or a physician is the feeling that he could help a desperate person heal and regain his health. The hospital, though charitable, felt that they gave enough so, I transferred the patient to a Rehabilitation Center which gave me reports on his condition from time to time. After a year, he could extend and flex his elbow and ten more months before he could move his wrist. Six more months before he started to move his fingers and another six months before he could use it. I saw him for the last time before he left the hospital and counseled him to be grateful to God for the gift He gave him to use his hand for a gainful employment and direct his life towards the good in himself and others.

I never saw him again, but you can see from writing his story that I never forgot him and I was rewarded by being able to salvage him and that I had the chance to learn and be experienced in trauma.

UNUSUAL TRAUMA NO ONE KNOWS HOW TO TREAT: JUDY'S STORY

Judy, a twenty-four-year-old ICU nurse, was leaving the hospital at the end of her evening shift. At the corner of the hospital, a young man knocked on her window and when she opened it, he shot her on the left side of her face with a BB gun before running away laughing. The poor nurse, shocked by what happened, returned to the emergency room, bleeding profusely from several holes in her face.

The ER doctor applied pressure dressings but these became drenched in blood. He typed and cross matched her for a blood transfusion and started alerting any of the surgeons on call. The general surgeon suggested she be referred to a plastic surgeon, who asked for her to be referred to an ear, nose and throat surgeon who in turn asked them to call a vascular surgeon. All of them refrained from taking responsibility stating that such trauma to the face was not in the realm of their expertise. Because the emergency room was filled with other patients, the ER doctor transferred her to ICU. Despite the pressure bandages, the patient continued to bleed profusely and the blood bank was running low. The head nurse did not know what to do with her co-worker bleeding massively, and no doctor was willing to accept the responsibility.

I was stirred awake early in the morning by a call from the head nurse, relating what had happened. In my sleepy state, I informed her I was not on call but before I hung up she urged, "Dr. S., I have a feeling if you don't come to save this nurse, nobody will." Such trust jarred me awake.

I immediately got dressed and headed straight to the hospital, along the way asking God for guidance. I had never dealt with a BB gun trauma at close range either, or read about it in any books. How to handle it?

By the time I got to the hospital, the idea hit me: if all the faucets in a building were dripping, it would be impractical to stop and repair each faucet. The most practical thing to do would be to shut off the building's main valve. With that hunch, I called the operating room and told them to bring the patient from ICU and asked the circulating nurse to prep the patient's neck for surgery. The anesthetist called me soon after, wondering if that was exactly what I meant. He was afraid I might still be half asleep.

The tension was evident just before I started the surgery. I could tell that much of the team was questioning my procedure. I made a left vertical incision in the left side of the neck and dissected the carotids. I asked the nurse for a hemoclip and used it to clip the external carotid artery from which all the branches that supply the face arise. To everyone's surprise, the bleeding stopped immediately. I looked at the resident assisting me, took out one or two pellets from her face and asked him to remove the remaining ones. There was no need to close the holes as this would cause healing scars.

When I called ICU later that day, the head nurse thanked and praised me, assuring me all the bleeding had stopped and the patient was stable. The patient wanted to speak to me personally but I asked her to rest up and I would see her in the afternoon. When I went to her all my colleagues were bragging saying, "We did it!" to the CEO of the hospital. But the head nurse was quick to interject, "He is the one who did it!" pointing at me. They all looked my way and could not figure out how I did it with the face and neck all wrapped up. Sarcastically, I told them, "I just ordered the bleeding to stop and it stopped." But I did so because not one of them wanted to come to her rescue. The holes eventually closed successfully, leaving no scars visible upon follow ups.

clip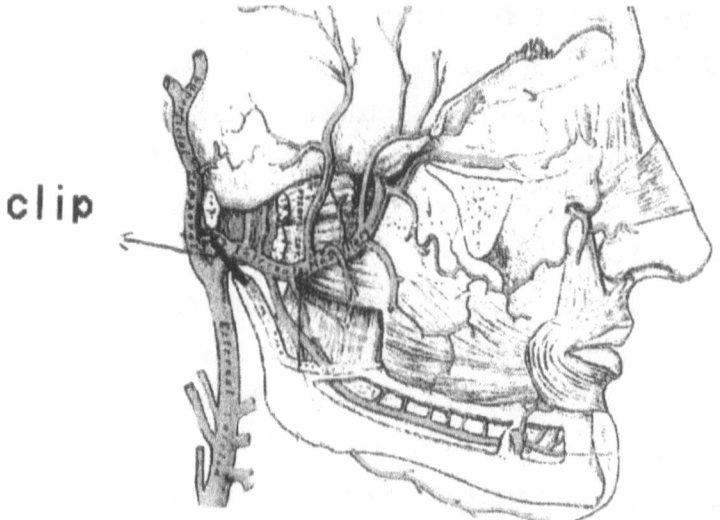

Figure 26: Clip on external carotid artery.

The arteries of the face and scalp are all branches of the external carotid artery. Occluding the external carotid artery would stop the bleeding from all its tributaries which were injured by the bullets and causing the hemorrhage. This is much easier than dissecting and occluding each vessel separately.

NOTE

The external carotid artery is a branch of the common carotid artery. The other branch is called the internal carotid artery which supplies the brain.

DRUNK DRIVING

INTRODUCTION

Driving under the influence is the most common cause of serious accidents and fatalities. Unfortunately, it occurs most commonly around the holidays when people unshackle dangerously and surrender to the leisure of the occasion.

Police can be somewhat of a deterrent, but nothing prohibits anyone from the pursuit of a desire to be free and happy at any cost. The worst judgment anyone can make at these times it to combine alcohol with other recreational drugs. This deprives the driver from his sense of reality and accurate estimation of the distances and speeds with which he is driving. At least one of the participants should remain sober to drive the others safely.

Another slip-up is falling asleep at the wheel. People drive longer hours than their concentration can reasonably handle and very often experience highway hypnosis. The best thing is to divide a long distance drive into segments where the driver rests in intervals and limits the number of hours he or she drives in one day.

CAR CRASH, MASSIVE HEMORRHAGE, BUT FROM WHERE: GARY'S STORY

It was Thanksgiving Day. I was full from the turkey dinner and settled down in front of the TV when I heard the telephone ring. It was the emergency room at Walther Memorial Hospital notifying me that a young man had been badly injured in a car crash, and his family physician was requesting that I, the chest surgeon on call, see him. His condition was grave with blood pouring from his right chest. He was not responsive and might die before I arrived. Luckily, that hospital was close to my home and, under a high adrenalin rush of my own, I was there within ten minutes. The young man in question had been brought in by ambulance, accompanied by the highway patrol. The officer informed me that the injured party had been driving at dangerously high speed when he hit a huge tree. The impact completely totaled his car, shattering the windows. He was found unconscious, leaning against the steering wheel, seat belt thankfully in place, but reeking of alcohol.

Routine examination by the emergency room physician and chest x-ray revealed a right hemothorax and low hemoglobin of 7gm, indicating massive internal bleeding. A chest tube was inserted on the right side and a huge amount of blood gushed from the chest and poured out of the tube. The family physician who had called me thought that there was definite injury to the lungs, heart, or main blood vessels in the chest.

Despite the massive bleeding, the X-Ray did not reveal an obvious source of this in the chest. The lungs were expanded, the heart was well delineated, and the mediastinum was normal in size. Contrary to what everyone else was thinking, I had a hunch the blood was coming from the abdomen through a hole in the diaphragm. With no time for hesitation I used my surgical intuition. The patient was exsanguinating terribly, and he needed an immediate laparotomy, as opposed to a thoracotomy.

To the surprise of the anesthesiologist and the team, I asked for the patient to be turned to a supine position, and for the abdomen to be quickly prepped and draped. As soon as I opened the abdomen with a midline incision, blood came gushing out. I expanded the incision to a full laparotomy and it was obvious that the spleen was torn apart. The left lobe of the liver was also torn, but before searching further, I surrounded the splenic pedicle and stapled it as a whole. Next, I stapled the bleeding part of the torn lobe of the liver and excised it. With the bleeding finally eased up I freed the spleen from its attachment to the kidney and colon by finger dissection. I removed it and packed the site it occupied with hot towels.

Further inspection revealed a tear in the diaphragm which allowed the blood to go into the chest, owing to the negative pressure created with each breath. I immediately repaired the tear in the right diaphragm. All sources of bleeding were tackled, including the spleen, liver, omentum, kidneys and the major blood vessels. It was a difficult surgery and took about eight hours and required about ten units of blood. With the cooperation of a competent anesthesiologist and keen determination on my part, not wanting to lose a patient of such youth, we were able to rescue him.

This successful procedure could not have happened had I unnecessarily opened the chest first. By that time the patient would have died during or even after surgery. Once the bleeding stopped, and with blood transfusions, the patient's blood pressure rose from 80/60 to 100/70 even before closing the abdomen back up. Once we were sure the abdomen was dry, it was closed in layers and a pressure bandage applied. The bleeding from the chest stopped, so we removed the chest tube the following morning.

Despite the apparent success of the surgery, there was still an immediate worry that the patient might suffer brain anoxia and never wake up. Therefore, it was decided to keep him on a respirator with a high output of oxygen.

It took eighteen agonizing hours before he woke up from the anesthesia and his alcohol intoxication. His folks were awake and by his side the entire time. They were devestated by what had happened to their son, and grateful that we were able to save him after all. When the young man finally woke up he was speechless. He gazed around at everyone and gradually became aware of his surroundings. Once he knew where he was and what happened, he sobbed. He seemed stricken by remorse, but his parents were not judgemental, and just hugged and kissed him. They were overjoyed to see him alive, and naturally, I was elated to have been able to save him.

We decided to try weaning him from the respirator, and under close observation we had him breathe humidified oxygen through the tracheal tube. On the fourth day he was stable enough to be extubated, he started ambulating. On the tenth day this young man was discharged and sent to a rehabilitation program which was required before he'd ever be allowed to drive again. Let's hope he learned his lesson.

NOTE

I think drinking should be outlawed on campuses and fraternities, and special rules should be enforced during holidays and fraternity parties. I hope such a story serves as a deterrent to others, and I wish that one of his fraternity brothers would have stopped him from driving in that condition. On all such occasions, there should be a few who remain sober to watch over the others. The penalty for driving under the influence should be an immediate revocation of the driver's license and not be returned before full rehabilitation.

As far as the seat belt is concern, I think it prevented him from flying out through the windshield, but it didn't prevent other injuries. In fact, I think because of the speed the patient was driving, and the gravity of the impact, the seat belt might have pressed on his abdomen and actually aggravated the injury. He was lucky it didn't rupture his intestines and cause peritonitis, which is known as seat belt injury. It did however rupture some of his abdominal blood vessels, causing severe hemorrhage. Even with all the benefits of the seat belts and air bags, they are no alternatives to cautious alert driving.

STAB AND GUNSHOT WOUNDS

A LAWYER'S BROTHER: JOE'S STORY

I was leaving the emergency room at Walther Memorial Hospital when I noticed a patient who was bleeding profusely from a chest tube on the right side of his chest. The patient was not receiving blood transfusions despite the massive amount of blood loss.

When I asked the ER doctor about this, he told me the blood bank did not have his compatible blood type nor was he able to get hold of the surgeon on call.

I told him he should do something quickly as the patient was at risk of exsanguinating to death. Suddenly, someone grabbed my arm from behind. It was the victim's brother who pleaded that I take his brother into surgery myself. He told me that he was a lawyer and the hospital will be responsible for keeping his brother in such a dangerous condition without help.

I told him that it was my instinct to help in such situations, but his being a lawyer is more intimidating than encouraging because he might sue me if his brother does not make it. "Moreover," I continued, "the lawyers took from us the Good Samaritan protection." He assured me that he would sign a consent which freed me from charges no matter what happens. I wasn't going to let him die without help anyway. I followed my surgical instinct and made up my mind to try to save him. At this stage the ER nurse informed me that the victim's blood pressure was barely audible and his pulse was thready and beyond counting. The patient was dying and something had to be done in a hurry. I immediately asked the blood bank to give him O type blood, and for the nurse to alert the surgery team to be ready as soon as possible.

The blood bank manager started to argue that it is "against the rules" to give him O type blood without testing for compatibility. I emphasized to him that the risk of his dying without getting blood replacement was much higher than the risk of a blood reaction. To reassure the blood bank I signed a paper that I'd take all the responsibility of giving him a general donor's blood (type O blood). In the meantime the patient was transferred to the operating room.

The anesthetist started a central venous line and started transfusing him with albumin and plasma until the first unit of blood arrived. It was transfused under pressure, followed by another unit of type O blood.

I made a fast scrub and quickly draped the bleeding side of the chest, performed an emergency thoracotomy between the 5th and 6th ribs and opened the space with a retractor to find the source of bleeding. After a quick inspection of the bleeding sites I discovered a tear in the lung and a hole in the heart and pericardium. I stapled the segment of the lung that was bruised by the knife.

I opened the pericardium wide to relive the tamponade and expose the hole in the heart. I put my left index finger on the hole in the heart from which the blood was spurting; this stopped the bleeding. The heart was pale and almost empty with a very weak systolic contraction. I waited until the coronary arteries started to fill up and the contractions became a bit stronger.

After transfusing a couple units or more of blood, the anesthesiologist told me that he could hear a faint blood pressure and pulse. I asked the nurse to hand me sutures with pledgets that do not cut through the heart muscle when tied. At that moment, I started to have hope that we could save the patient's life, and began ligating the sutures one by one. It was comforting when I removed my finger and the stitches closed the wound. The bleeding stopped completely.

A sense of relaxation came over me as soon as his blood pressure steadily rose with the increased transfusion. I prayed with all my power that he didn't suffer brain or kidney damage from the hypotension, and consoled myself for being in the proper place at the right time, passing through the emergency room by accident, and the fact that we didn't waste a minute to stabilize him.

The chest was closed in the usual manner leaving a chest tube to drain any extra fluids or air. Dressings were applied and the patient was turned on his back. My prayers were answered when the anesthesiologist told me his pupils were reactive, a sign that the brain was intact. I looked at the urine bag and was elated to see 400 cc's of clear urine, meaning his kidneys functioned during the operation.

We transferred the patient to the ICU and gave a cautious report to his brother, telling him the operation had gone well, that the bleeding had stopped and that his brain and kidneys seemed to be undamaged. The next day the patient awoke and, now breathing on his own, was extubated by the anesthesiologist. Being fully conscious, the patient told us he was a math teacher named Joseph F., and explained what had transpired.

He owned rental apartments and had gotten into an argument with a tenant over late payment. Things heated up and the tenant pulled a knife and stabbed him.

"Don't worry," I said. "Your brother, the lawyer, will take care of the guy who stabbed you."

It's always a good feeling when a patient gets out of bed, ambulates, and gradually eats a normal diet. His pain subsided, and every time I went to see him his fiancé was romantically by his side.

After ten days the patient was discharged, though he continued to come monthly for three more months in order to monitor his high blood pressure until it stabilized. His post-operative x-rays remained normal as well as his pulmonary function test. Some years later, I asked him about his brother, the lawyer, and learned that he had been promoted to Appellate Court Judge.

NOTE

1. Routine orders should be overruled when the patient's condition demands otherwise. If I didn't insist that the patient be given type-O blood at the proper time, the patient was moments away from dying of exsanguination.

2. To save such a patient, it's vital that the patient be in the proper place at the proper time with the proper doctor to handle his condition.

3. The Good Samaritan Law should be reinstated wherein health practitioners can administer help in good faith to anyone in need without risk of being sued. This law rescinded creates apathy in people, medical practitioners or otherwise, whereby they pass by a person in need of help and keep going, afraid their involvement might lead to a law suit.

4. Stab and gunshot wounds are very common among gangs in the big cities and I pray for a time when we can teach people love and civility instead of violence. When I was coaching residents in Mt. Sinai Hospital, not a night passed without being called in to guide them through the treatment of such serious traumas.

5. Guns should be licensed and owned by responsible people who need to protect themselves.

MACHINE-GUNNED: JOSE'S STORY

Jose P. owned a Spanish nightclub in downtown Chicago which brought him generous revenue. One night, an armed gang broke into the club and machine gunned him and his brother. He was in critical condition, bleeding from his chest and abdomen. He had almost no pulse and was barely breathing. His brother had already been declared DOA.

By the time Jose arrived at the ER he didn't have a pulse and I had to call a "code blue" during which he was stripped of his clothes to inspect every wound in his body. Bullet holes riddled his chest and abdomen with numerous exit wounds on his back. The team took quick x-rays of his chest, abdomen, and pelvis, while a blood bank team took a blood sample for CBC and blood matching. The results arrived quickly: one bullet was lodged next to his left thoracic spine and he had a left hemothorax. A tube was inserted in the chest to drain it. The CBC was very low and the blood bank sent us two units to start with but we had to find more blood of his type through the Red Cross or neighboring hospitals. His abdominal x-ray revealed gas under the diaphragm, which indicated perforations along the gastro-intestinal tract. A Foley catheter revealed blood-stained urine.

Wasting no time, I ordered the patient be transferred immediately to the operating room. The OR staff had previously been alerted and as I quickly scrubbed, the anesthetist intubated the patient and put him to sleep. Then I pondered: should I open his chest or abdomen first?

193

My intuition was that the injury in his abdomen was more life threatening than the one to his chest.

There were several reasons I arrived at this conclusion. The bleeding from his chest tube was not massive, but perforation of the bowel into his peritoneum can cause peritonitis, which can cause deadly infection. The pattern of the bullet entries and exits suggested they had passed through the spleen, and possibly the vena cava, aorta, bladder and bowel. Upon performing the midline laparotomy, a gush of blood and other fluids poured out. I quickly vacuumed these fluids and the blood clots, and tamponaded others with warm sponges. The spleen and abdominal vessels were bleeding profusely, the liver and gallbladder leaking bile, and the holes in the intestine were excreting digestive fluids.

In order to stop the hemorrhaging, I decided to perform a quick splenectomy. I began by stapling the artery and vein at its hilum, freeing it from its connection with the colon and kidney, and removed it. I then turned my focus to the abdominal vessels. As soon as the pressure was removed from them, a gush of blood came out the right side. It looked like venous blood and I knew then that it was coming from a hole in the vena cava and not the aorta, because the aorta is on the left side, and blood pouring from it would have been much more red and oxygenated.

To stop the hemorrhaging from the vena cava, I had to separate it from the aorta, and then I applied a straight atraumatic clamp below the hole. This stopped the bleeding except for a trickle of back bleeding. I replaced the straight clamp with a J clamp over the hole that occluded half of the lumen and allowed venous return to the patent half. Thus I could suture the hole over the J clamp and then remove it. This allowed total venous return to the heart. With the bleeding stopped and the heart filled with blood, it started to beat slowly. At this point the blood pressure could be measured at 60 systolic, which is very low, but better than nothing.

Attention was then directed to the holes in the bowel. After removing the towels, I found three holes in the jejunum and a tear in the mesentery. Rather than closing the holes, I removed that segment and connected the bowels with an end to end anastomosis.

For the tear in his gallbladder, I preferred to do a fast cholecystectomy: clip the cystic artery and the cystic duct and remove the gallbladder from its liver bed with my fingers.

The urinary bladder did not pose any immediate danger, due to the urine being drained by the catheter, and thus it was left for last. But because it was intra-peritoneal, the bladder was closed in two layers and covered with omentum.

Before closing the abdomen, I found a bullet hole in the left diaphragm through which I felt the heart and the pericardium pulsating strongly, with no blood in the pericardium. The lung seemed well expanded in spite of a little hole in the surface of the lower lobe which was not bleeding; the heat from the bullet had cauterized it. I decided to leave the chest tube alone and monitor the lung by daily x-rays. The bullet, which remained close to the vertebral column, had not fully penetrated or caused spinal cord injury. I could feel it with my index finger and with a little pressure and manipulation I was able to free it from the bone and extract it through the hole in the diaphragm. The diaphragm was then closed in two layers and the abdomen closed by retention sutures. Blood loss was measured at about three liters and replacement required eight units of blood. Urine output after suturing the bladder was 150cc's.

With more transfusions the anesthesiologist was able to stabilize his blood pressure first at 90/70 to 100/60. To avoid overload from excessive transfusions, a central venous catheter was inserted to measure the pressure of blood in the heart chambers. Fortunately the kidneys hadn't shut down, as witnessed by the flow of urine in the bag.

We began the operation at 5:30 PM and it was now 12:30 AM. Needless to say, I was exhausted. After closing the abdomen, I waited in the operating room lounge with a cup of coffee. An hour later, the anesthesiologist informed me the patient was stable. We transferred him to ICU and connected him to a respirator. There was no drainage from the chest tube. We were not sure whether he had sustained any brain damage, but our concerns were alleviated the next day when he opened his eyes and seemed alert.

After forty-eight hours, we were able to wean the patient from the respirator and remove the chest tube. As soon as the bowel sounds returned, we started him on a progressive diet and ordered ambulation and eventually physical therapy.

Upon his arrival to the ER, with his brother killed by similar wounds, nobody thought he'd make it. But if there's a beam of hope to save a fatally wounded patient, the surgeon should have the courage and self-confidence to follow that beam.

GAMES OF THE HEART

After I finished the operation, an attractive, middle aged woman was waiting for me. She told me she was the patient's wife, and asked how the operation went.

"Thank God he's alive," I immediately responded. "We have every reason to believe he'll make it, but we'll continue to closely monitor him in the ICU."

The next morning, I received a long distance call in my office from Mexico. The caller said she was Jose's wife.

"When did you go to Mexico?" I asked. "I talked to you right after surgery." "We never met," she replied. "I'm Jose's real wife, mother of his children. That woman you met in Chicago is his girlfriend."

Placed in the middle of an awkward situation, I focused solely on the medical situation and told her that Jose was alive and stable.

Within two weeks of being admitted at the brink of death, Jose was well enough to go home. He scheduled a follow up visit with me. After I took care of him he asked if I could see his girlfriend who exhibited some small health problems. She eventually became my patient and came with him whenever he had an appointment. As circumstances would have it, it was through her that I stumbled onto my next memorable case.

SEQUENTIALITY: THE SEQUENCE OF EVENTS

It's strange how one incident often leads to another, and yet another might unexpectedly lead you in a direction you never expected. The following is an example of that.

While waiting for me during one of her appointments, Jose's girlfriend, Sylvia, was immersed in a book she was reading, when I entered the examining room. By curiosity I asked her about it, and she replied, "It's a Unity book."

"What is Unity?" I asked.

"It's a kind of mystic religion that teaches how the inside controls the outside, and not to let the outside control you."

"Where do you practice this concept?" I asked.

She told me there was a Unity Church in the neighboring suburb of Oak Park which was very close to my office. I jotted down the address and phone number.

Upon finding the address, I came to a home that looked less like a church and more like a huge mansion or club. I gathered my courage and rang the bell. Inside, I was fascinated to see a huge library with all kinds of books including, eastern and western philosophy, self-improvement books, etc.

The secretary at the front desk informed me there was Bible class every Wednesday at 7:00 PM and a formal lesson at 9:00 AM and 11:00 AM on Sundays.

I went the following Wednesday, curious to attend this Bible class. The speaker piqued my interest as his lessons focused on applying the stories from the Bible in our own lives.

One of the things he said that stuck in my mind was, *"Greater is He inside you than in the whole world."*

After the lecture, as I was browsing through the library, the speaker came over and introduced himself as Reverend Billings. I thanked him for the interesting lecture and told him I would come more often. He told me that unfortunately he was going to retire next month because his doctor had told him he was experiencing heart failure and irregular heartbeats, and this would not mesh with the pressure of leading a large congregational church by himself.

He asked me for a second opinion. I gave him the address of my office and told him to call for an appointment.

TO MINISTER OR RETIRE? A REVEREND'S STORY

Reverend Billings came to my office to tell me he'd found a new minister to replace him. His major complaint was irregular heartbeats, burning pain under his sternum, especially after heavy meals, and shortness of breath when walking long distances. It's true that these symptoms are suggestive of angina, and the shortness of breath could be from pulmonary congestion.

His chest x-ray revealed a big mass attached to his left ventricle which read as severe cardiomegaly or ventricular aneurysm. His lungs were clear and his EKG was normal and did not show a sign of left ventricular enlargement or an infarct. I performed a treadmill test on him and he was able to complete the test, so there was a big discrepancy between his symptoms, chest x-rays, EKG, and treadmill findings. How could I explain such a discrepancy? After thinking about it for a while, I ordered a CT scan of the chest. This solved the problem. The mass attached to his heart turned out to be a huge para-esophageal hiatal hernia. That is why I tell my fellow doctors to always **look outside the box** and use their own intuition.

Here's the problem: whenever he ate, the hiatal hernia expanded and pressed on his heart causing his heart to beat irregularly and interfere with his diaphragmatic breathing, causing shortness of breath.

This is a prime example of how a misdiagnosis can affect people's lives. The x-ray report had originally been interpreted as cardiomegaly, which misled his doctor into interpreting his symptoms as angina and heart failure, which prompted him to urge Reverend Billings to retire unnecessarily early.

After fixing his hiatal hernia, and closing the hole in his diaphragm, the symptoms disappeared. He was able to resume preaching, and to this day he's active in his church.

If I happened to be present at one of his sermons, he always made it a point to express his gratitude and tell the congregation that without my help he wouldn't be there.

I am the one who needs to thank him, however, as it was a true blessing for me to know him. He's a gifted, loving speaker whose sermons are inspiring, and they connect with the entire congregation.

Ten years later his blood test revealed a high PSA. I took a biopsy from his prostate and it turned out to be cancer. Under the guidance of my oncologist friend, I treated him with a pill and a monthly injection (hormonal therapy). It was a pleasure to have him as my patient until the day I retired. Until today, he still remains one of my best friends.

Wrong diagnoses are a big problem in medical practice. Not only does the patient not receive the effective treatment for his actual condition, but he acquires the symptoms of the wrong diagnosis through the mind-body effect, which can unnecessarily devastate a person's life.

I cannot help but put the following poem within this context, because Reverend Billings gave it to me. I find it very meaningful.

Each Life Affects Another's

We may not always realize that everything we do
Affects not only our lives but touches others, too.
For a little bit of thoughtfulness that shows someone you care,
Creates a ray of sunshine for both of you to share—Yes,
every time you offer someone a helping hand,
Every time you show a friend you care and understand,
Every time you have a kind and gentle word to give
You help someone find beauty in this precious life we live,
For happiness brings happiness, and loving ways bring love,
And giving is the treasure that contentment is made of.

—Amanda Bradley

SECTION VII

ERRORS IN DIAGNOSIS AND THEIR CONSEQUENCES

ERRORS IN DIAGNOSIS

INTRODUCTION

In the medical profession's zeal for perfection, quite a bit of auto-criticism has been published by doctors. Dr. Jerome Groopman, a professor of medicine at Harvard, aimed at correcting this problem in his book *How Doctors Think*. The premise of his book is to explain the type of thinking that results in a wrong diagnosis.

Several factors may lead to a wrong diagnosis. Many times, doctors depend on an incorrect x-ray interpretation. Other times, doctors *don't listen* to the patient's *detailed history*. Another habit is stereotyping symptoms. Doctors automatically think of the *most common diagnosis*, while the patient might actually suffer a rare condition.

The trouble with wrong diagnosis is that the patient receives the wrong treatment for the wrong condition which could lead to serious complications. And if the person believes that he suffers the wrong disease, his body acquires its symptoms (Mind/body effect).

Correcting a wrong diagnosis needs a lot of courage and self-confidence to oppose other doctors' opinions. Apart from intuition he needs a lot of research and conviction to support his corrected diagnosis.

The correct diagnosis will direct the doctor and the patient to the right treatment that leads to a cure, and abolishes a lot of suffering and complications.

In the next few pages, three patients will be presented as examples of wrong diagnosis and its correction, in addition to Reverend Billings' misdiagnosis in the last section.

207

IS IT TRULY MS? LIDA'S STORY

Fifteen years ago I went to Miami on a short vacation to escape the cold weather of Chicago. I was visiting with my cousin Jacki when she introduced me to her friend Lida whom she said was seriously ill and asked if I could help her.

Lida was a charming young lady who appeared less than her real age. She looked sad and depressed, and her husband said she couldn't get out of bed and was consistently weak and tired. Lida's eyes drooped and she was lethargic. She had been diagnosed with Multiple Sclerosis (MS) which is a chronic slowly progressive disease of the central nervous system that causes a wasting of the nerve sheaths of the spinal cord. It typically progresses in fits that cause crippling degenerative symptoms in almost every system in the body. It's characterized by remissions and intermissions that overall, result in weakness and paralysis in the arms and legs and a shorter, compromised life.

Lida showed me numerous CT scans and MRI's which had resulted in that diagnosis, which was confirmed by professors and chairmen of radiology and neurology in Israel and the USA as Multiple Sclerosis (MS).

Although I am not a radiologist, I had a feeling that this diagnosis might be wrong. In her desperation, the patient went from doctor to doctor, including Chinese and acupuncture specialists, who all confirmed the diagnosis of MS. I had a hunch that something was wrong and asked her to visit me in my office in Chicago.

She didn't seem to believe me when I told her there might be a mistake in the diagnosis. She could not believe that a surgeon like me is more knowledgeable about her neurological condition than those professors of neurology and others in two advanced countries. I went back to Chicago with the agreement Lida would visit me for a detailed history and an accurate physical examination.

In the meantime, I started to review the literature and the x-ray findings in MS, and compared them to the ones she gave me. I came to know that the characteristic CT scan findings in MS are white spots in the back of the brain which were lacking in all the patients' CT and MRI scans.

When they came to visit me two weeks later, I started taking an accurate history. I asked her if the symptoms came gradually and persistently or in attacks. She told me they were gradual and progressed consistently, and became permanent since the diagnosis a year earlier. This was not typical of MS, which comes in attacks with intervals of intermissions and remissions.

With the absence of the white spots on her x-rays, I became convinced the diagnosis was wrong. The symptoms didn't correlate with the x-ray reports, and those professors followed what I think was an erroneous radiological diagnosis, hence my advice to always **Look outside the box.** There's no substitute for a patient's accurate history and thorough physical examination.

"What could it be?" I wondered. The idea that came to my mind over and over again was that her symptoms were typical of chronic fatigue syndrome (CFS). Blood tests showed her thyroid functions, (a common cause for CFS), were normal. I also asked her about any yeast in her environment. She didn't have rugs in her bathroom which could store yeast spores, nor did she have any changes in her house suggestive of yeast infestation. There was also no history suggestive of Epstein Virus symptomatology.

When I was close to exhausting all the conditions that cause CFS, her friend suddenly volunteered that the patient had a bowel movement only once every ten days.

That meant the patient had yeast spores residing in her bowels or stools obstructing her bowels causing a leaky bowel syndrome and/or a toxic colon.

Thanks to my habitual use of intuition, a previous, gloomy diagnosis turned into a simpler one to explain and cure her symptoms.

The round spots in this brain scan are characteristic of

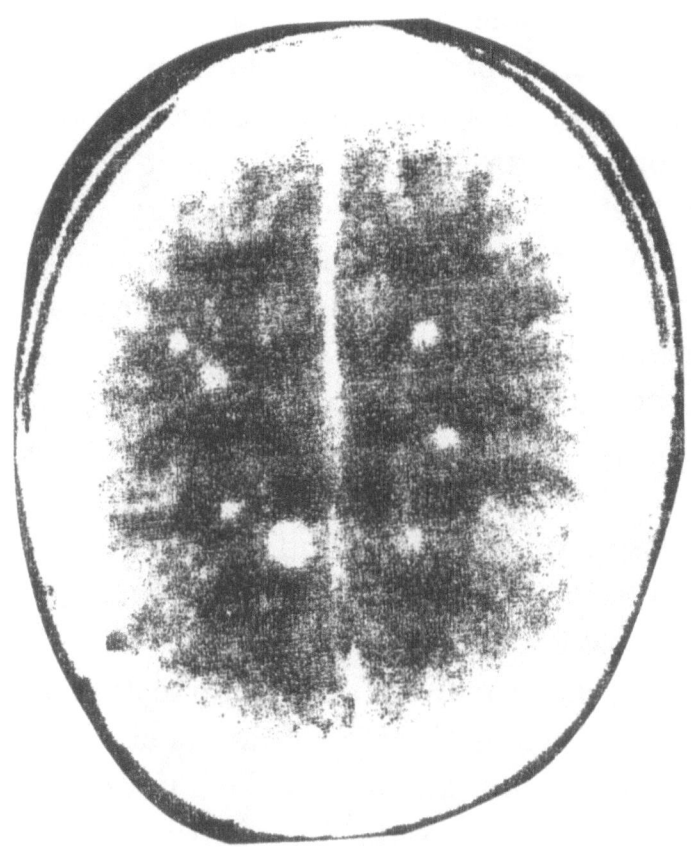

Figure 27: Brain Scan

Multiple Sclerosis. Their absence on her CT scans made me suspicious of the radiologist's and neurologist's' diagnosis. That's what looking outside of the box means.

The correct treatment was to start with an immediate bowel cleansing, followed by a diet rich in fiber, and drinking plenty of clean water to regulate her bowels. As for medicines, she was given Flagyl 2000mg daily for two weeks, megavitamins, minerals, probiotics and basic antioxidants. I counseled her regarding stress and fatigue and explained to her how chronic constipation can lead to

211

toxic colon syndrome.

A few days after this vigorous treatment, she returned to Florida with a smile on her face. She started to get less tired than before. I kept communicating with her and she reported that she had regained her energy and started to look for a job to help with her family's finances. A few years later, she got a permanent job as administrative assistant at the Holocaust Museum in Miami Beach. She's happy and enthusiastic about her work and exudes love and compassion to all those who visit the museum, especially the Holocaust survivors.

Lida was very appreciative and became a close friend to me and my wife. Today, she and her husband, Nissan, are two of our closest friends.

The next year we planned once again to return to Miami to escape the freezing Chicago winter. Lida offered to pick us up from the airport and we called her upon our 9:00 PM arrival. She arrived dressed in a long formal dress. When I told her how pretty she looked, she said she was at a party, but had excused herself to pick us up and would return to the party after dropping us off.

This act was a display of true friendship, compassion, and gratitude. Our friendship with Lida and her husband grew, especially after my retirement in Florida. They became as family members and we always share happy occasions, feasts and gatherings.

NO MORE PNEUMONIA OR CHF: SOLEDAD'S STORY

Soledad was a fifty-year-old Spanish lady who came to my office accompanied by two intelligent, beautiful daughters who translated for her. She was referred to me by a family practitioner with a diagnosis of angina and repeated pneumonias which had her in and out of hospitals every other month. When I saw her, she looked well-nourished and smiled consistently and Nitroglycerin relieved her angina.

She said she was experiencing retrosternal pain which happened more often when she drank hot or cold drinks, and sometimes after meals, but not radiating to the arm or neck. These symptoms were usually relieved by Nitroglycerin. The pain had started three years back and was moderate in nature. She suffered from a continuous cough, sometimes productive and sometimes dry which was worse at night, and sometimes woke her up.

On examination, the patient had bilateral rales and wheezing on the upper and middle parts of her right lung. On auscultation, a soft murmur was heard in the aortic area, but other heart sounds were normal. Her blood pressure was 130/90, temperature normal, abdomen soft with no particular area of tenderness, and there was no ankle edema.

A chest x-ray revealed a middle lobe syndrome and diffuse areas of atelectasis, and her EKG was border line. She was scheduled for a treadmill and was able to stay on it for only six minutes. Her blood pressure went up to what was expected but we could not increase the speed to a level consistent with the protocol of the machine. Her legs became tired, so we stopped the treadmill. Her 2D Echo (sound waves of the heart) showed a normal size heart with normal wall motions.

At that time her symptoms had been interpreted as coronary artery disease with pulmonary venous congestion. What was puzzling to me was that she had this for three years, but her heart condition had not worsened. Also, angina did not come after effort, such as walking or climbing stairs. It came only after eating or drinking hot or cold liquids. That made me uncomfortable with her diagnosis.

I asked her about heartburn, but she could not understand, and confused it with chest pain. So, I asked her if she ever used Tums or Rolaids to relieve the pain. Her answer was that she used them frequently and they helped. I called her family physician to ask if she had had an upper GI and he said that she did, but it was normal.

In cases like these, where anything is possible, a doctor has to trust his intuition. I suspected that she had a disorder in the function of her esophagus and sent her for *esophageal motility studies*. These revealed esophageal spasm (tightening in the lower esophagus) and high pressure as indicated in *manometry*. (A manometer is a tool to measure pressure, in her case, in the lower esophagus.) This could cause the acid and food in the stomach to regurgitate. Because the pressure is higher in the lower esophagus, the food and acid cannot reach the stomach as usual. The food and acid might then fill the esophagus, spill over and be aspirated into the trachea and lungs, causing inflammation and infection. When lung infections are so repetitious and there are no indications of COPD, we must always think about esophageal dysfunction as the cause. This could cause severe reflux that awakens the patient to coughing attacks.

The retrosternal pain associated with this disease, and its being relieved by Nitroglycerin, could mislead a physician to think it's angina pectoris. But the fact that Soledad's symptoms were exacerbated by hot or cold drinks, and localized in the lower retro-sternum, without radiating to the arm or neck, confirmed an esophageal origin, not cardiac.

A bronchoscopy showed plaques of mucous, especially in the area of the right middle lobe, and the patient felt well after aspiration and bronchial washings. The results of the washings showed there were no malignant cells, but many leucocytes. These could be caused by reflux and aspiration.

I admitted her to Thorek Hospital and scheduled her for a procedure called *esophageal myotomy* or *Heller's Operation*. This means cutting the muscles of the lower third of the esophagus that causes the spasm, allowing the mucous membrane to bulge out. This takes away the excessive muscular spasm and allows food and drugs to pass smoothly from the esophagus to the stomach without pain or regurgitation. This procedure stopped the pain, aspiration, and infections in the lungs completely.

See figure 28 on page 216 for an image of the operation.

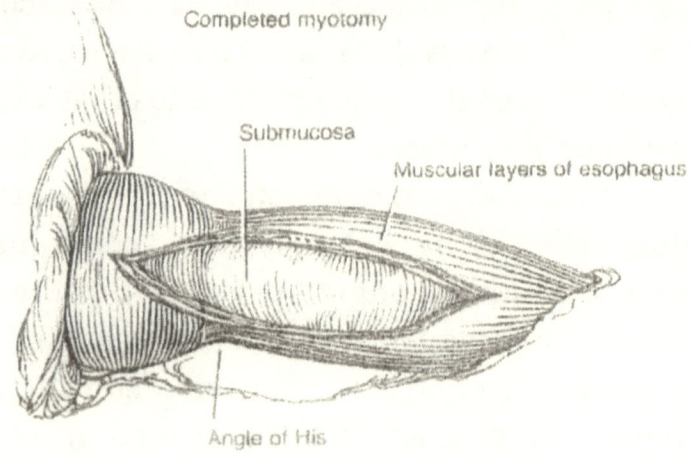

Completed myotomy

Submucosa

Muscular layers of esophagus

Angle of His

Figure 28: Cutting the muscle of the lower esophagus, which caused the spasm and restrosternal pain.

Post-operatively, she was placed on intravenous antibiotics which resulted in a decrease in coughing. Her daughters brought her for weekly visits and were gratified that the coughing had stopped completely. I put her on a soft diet, and to my gratification, she could eventually drink anything, hot or cold, without feeling pain. A year later, she moved to Florida with one of her newly married daughters. Seven years later, she came to Chicago and complained of arthritis in her knees. She expressed her gratitude for stopping her cough and curing her so-called pneumonia permanently.

CONCLUSION

I did not consider the angina a heart condition, nor the lung changes as emphysema or pulmonary venous congestion, as reported by the radiologist. I ***looked out of the box.***

Why did I diagnose it as this dysfunctional disease of the esophagus?

1. Absence of cardiac science by the EKG, the treadmill, and the ultrasound of the heart, which revealed good size and function of the heart.
2. The use of the antacids.
3. The pain was confined to the lower esophagus and not the whole esophagus, neck and shoulders, as in cardiac ischemia.

The concluding point in the defense of the diagnosis is that in both conditions, the pain is relieved by nitroglycerin. This needs accurate **listening** to the patient. As Dr. William Osler said, *"If you listen carefully to the patient, he will give you the diagnosis."*

ANKLE SWELLING: DOROTHY'S STORY

The following case is really ridiculous because of the foolish misdiagnosis that was made. The patient was sent to me by a podiatrist, and she had been referred to him by a general practitioner for surgery because of persistent pain and swelling in her ankles. The podiatrist sent the patient to me for medical clearance.

Dorothy was a forty-five-year-old bookkeeper who complained of arthritis in the knees and ankles. On examination, she had congested neck veins, basal rales, an enlarged liver, swelling in her legs and feet, and a murmur. Her EKG revealed left ventricular enlargement. 2D Echocardiogram revealed right and left ventricular dilatation, left atrial enlargement, moderate-mild mitral insufficiency and decreased left ventricular wall motion. Her history revealed dyspnea on exertion and orthopnea, meaning that she had to sleep on two or three pillows. She also complained of easy fatigability, lack of appetite and palpitations.

The diagnosis was obvious of CHF (Congestive Heart Failure), causing the swelling in her ankles. Her treatment consisted of digitalization and diuresis. Within a week her leg swelling disappeared and, of course, ankle surgery was cancelled once I informed her podiatrist of my findings.

The patient refused to go back to her referring physician. When I asked him how he missed the diagnosis of CHF, which was obvious, he said, "The x-ray report said that the heart was borderline enlarged and lungs were clear."

CONCLUSIONS

The first mistake was on behalf of the doctor when he relied more on the negative x-ray report than the history and physical exam. The second mistake was by the patient who put more emphasis on her ankle swelling and minimized the degree of her dyspnea. Dorothy suffered for months with her leg swelling and pain, shortness of breath, palpitations and lack of energy—all of which had been misdiagnosed as arthritis of the ankle!

Hence the theme and advice throughout the book is to **look outside the box**. If the x-ray diagnosis conflicts with the symptoms, look deeper for the right diagnosis.

SECTION VIII

THE VALUE OF ATTITUDE

THE VALUE OF ATTITUDE

INTRODUCTION

Mayer's story is a great example of how our attitude and mindset affect our bodies. I have witnessed many patients who have been given devastating prognoses, but refused to accept it and surprised their doctors. The will to live is hard to defeat. I've seen people delay their moment of death when expecting a relative or dear one to arrive.

Like Mayer, possessing a strong mind, cheerful spirit, faith, hope and acceptance with a positive mental attitude, we can overcome any disaster in our lives. In the words of Isaiah, *"He who overcomes shall have everything; he will be my son and I will be his Father."* (Revelations 21:7.)

THE SMILE THAT KEEPS HIM ALIVE: MAYER'S STORY

Mayer is my cousin, a man who defied the odds. His courage and attitude keep him alive and resilient. He currently lives in London and is about my age. In fact, we grew up together, and he'll always be my best friend and hero.

At the age of twenty-six, he started complaining to me of weakness in one of his legs. A few months later, he was limping. In time, the other leg started to become weak and next thing we knew, he was confined to a wheelchair. A muscle biopsy revealed he suffered from Muscular Dystrophy, an autoimmune disease which to date is incurable. It usually kills its victims within five years. Mayer, however, is a unique being, and despite everything, continues to be cheerful. He enjoys a good joke, has a wonderful sense of humor, and loves to laugh. He never complained, even when the dystrophy spread to his arms, hands and fingers and rendered him incapable of holding a fork. All he can do is sip with a straw.

The problem with muscular dystrophy is that sooner or later it affects the respiratory muscles and kills its victims from apnea. Mayer accepts his disability cheerfully, without self-pity. If he were seated next to you, you wouldn't know he's sick. You'd think he was a normal person confined to a wheelchair.

In 1990, he hosted his daughter's wedding. Friends and family came from all corners of the world, and he thanked God for blessing him with a good and loyal wife who arranged his elegant attire and transportation. Mayer always looks at the bright side of life.

Last year we visited him in London. His wife drove, and he was seated in the passenger seat. He was very lucid in giving her directions for the closest way to the restaurant where they had made reservations for that night. He had a very devoted aide sitting in the back seat.

As soon as we arrived, the aide brought his wheelchair from the trunk and practically carried him from the car to the chair and wheeled him to the restaurant. He sat beside him and fed him.

As we dined, we reminisced about our childhood, our school memories, and his business. He was honored by the Queen of England with the award for Export Achievement. He sends light to every dark house in Africa with his generator factory in London, using a grant from the World Health Organization. He has become a popular and blessed figure in Nigeria, but he never brags when asked about how he came to receive such an honor from the Queen. He credits it to his proud son who reported the achievements to the Secretary of Trade. (See Figure 29 on page 226.)

**Figure 29: My cousin Mayer Godsi (middle) and his son
Solomon (right), receiving the Certificate of Excellence for Export
Achievement, granted to Mason Overseas LTD.
Field Marshal Lord Brummel (left) delivers the decoration and
certificate on behalf of Queen Elizabeth II.**

It is now fifty years since he was afflicted with a disease that could have killed him within five. Very casually, in the course of a conversation, he told me that last year he had a coronary bypass. I was surprised at how he survived the operation. To him, it was just a minor event that occurred during his lifetime. It was amazing to me when his wife described the ease with which he accepted the operation, (without fear or complaint), and the smooth recovery followed by his return to work. He treated it more like a picnic than a serious operation.

Whenever there is a wedding, Bar Mitzvah, or any celebration, locally or abroad, he will always attend, along with his wife and helper.

I really consider Mayer the hero of our family. He's an example of a man who accepts and overcomes his disability without complaining or feeling sorry for himself. He conducts multimillion dollar businesses with an alert brain and cheerful soul. When I call him I tell him, "Mayer, keep that smile because it keeps you alive."

He enjoys traveling and when he comes to America, always stops at the Taj Mahal in Atlantic City to indulge in some gambling.

I include Mayer's story to show everyone that a cheerful, positive attitude, even in the face of a fatal illness, can defy the odds and prolong your life.

SECTION IX

COMPRESSION
SYNDROMES

COMPRESSION SYNDROMES

INTRODUCTION

Compression syndromes are conditions that cause numbness and/or pain in a limb. They are often misdiagnosed and instead of compression, they are referred to as peripheral neuritis, diabetic neuritis, ischemia pain, disc herniation, neural narrowing in the spine, etc. When none of these is proven by tests or x-rays, they are said to be psychosomatic, neurotic and hypochondriac, and the patient is sometimes referred to a psychologist or psychiatrist unnecessarily.

Two patients, with **two types of compression** will be discussed in this chapter:

1. Leg compression, with an unusual case description.

2. Arm compression, the common method of treatment, and the author's modification.

LEG COMPRESSION

NOBODY BELIEVES ME WHEN I SAY MY LEGS HURT: DIANE'S STORY

Diane was a twenty-six-year-old American Airlines stewardess. She was referred to me by an American Airline employee by the name of Jerry Hill who was a patient of mine for many years. Diane gave me a story of pain in her legs when she stands for long periods of time, and asked to be assigned to shorter flights. She said that she had undergone a complete workup by the airline physician, also a vascular surgeon, a neurologist, and a spine and orthopedic specialist. She said they did arteriograms, venograms, Doppler studies, x-rays of the spine, CT of the hip joints, and they all came back normal.

All those doctors concluded that her condition was "psychosomatic" and finally sent her to a psychiatrist for evaluation. She said that she was skeptical about seeing any more physicians, except that Jerry told her, "Salvador is the only one who can find what is wrong with you when nobody else can explain your pain."

ACCURATE HISTORY

She gave a history of falling from a bike a few years before her employment, and ever since then she had pain when she stands or walks vigorously. The longer she stood, the worse the pain. It was neither claudication nor sciatica.

232

When asked to point out the exact location of the pain, she indicated the lateral front part of the leg. She said that analgesics like Aspirin or Tylenol were not very effective, and Plavix, given to her by one of the airline doctors for a few months, was of no benefit. Added to that, she said she enjoyed her job as a stewardess and had no desire to change her profession.

ON EXAMINATION

Her vital signs, blood pressure, pulmonary function test and blood tests were within normal limits. Her leg pulses were palpable. Leg elevation did not cause any pain. There were no varicose veins in her legs. She was tall, slim and attractive and did not seem psychologically disturbed in any way. The only thing which stuck in my mind was that her pain was confined to the anterior lateral segment of the tibia corresponding to the anterior compartment of the leg.

I measured the circumference of her legs with a tape and then put her on a treadmill with high speed and a slope. After about ten minutes she said the pain started to come, and a few minutes later she said that it was severe and asked to stop the test. I again measured the circumference of the legs after the treadmill and I found there was about half an inch increase.

MY DIAGNOSIS

I had the impression that on standing a long time the anterior tibial muscles got swollen, and the compression between the bone and the fascia surrounding the muscle was the cause of her pain. Added to that, the injury of her falling from the bike, and some puffiness in that part of the leg, confirmed my suspicions.

THE TREATMENT

I explained my thoughts to Diane and suggested to her that a Fasciectomy (i.e. cutting through the membrane that surrounded the muscle) could free the swollen muscle from the compression and give it the chance to expand freely and without tightness.

She asked about the anesthesia and if she should go to the hospital for it. I told her it could be done under local anesthesia through a small incision below her knee. She could get up and walk after it with maybe only a little rest.

She agreed and wanted to try one leg on her day off and if it benefited her, she would come back for the other leg.

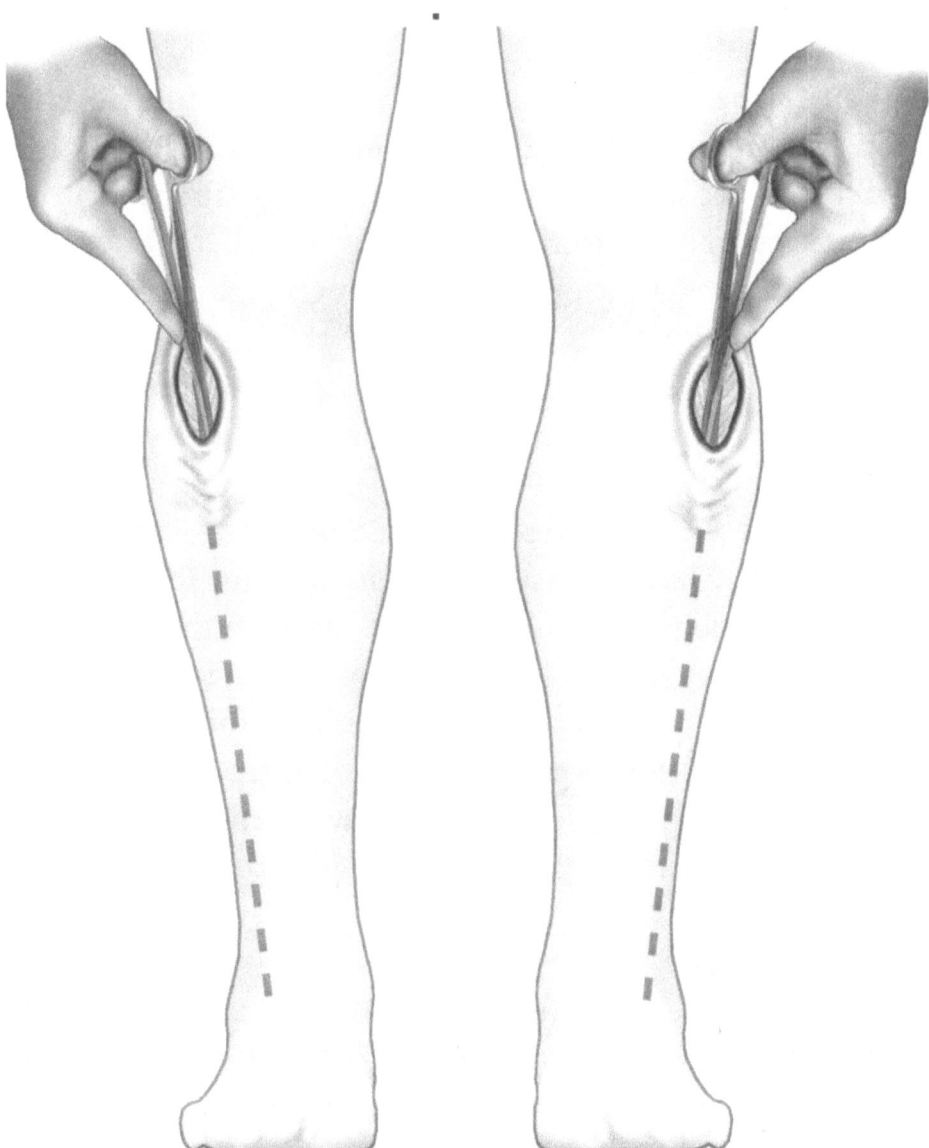

Figure 30: Anterior chamber fasciectomy.
Anterior compartment release.

THE OPERATION

Under local anesthesia, I made a small incision an inch lower than her tibial tuberosity and dissected under the skin till I saw the shining fibers of the membrane in question.

With a long scissors I cut through it as far as the scissors could go, about two thirds of the leg. The incision was closed with staples and a small dressing was applied over it. Jerry drove her home and she resumed flying after the weekend.

A week later Diane came for removal of the staples and she reported that, miraculously, she did not feel pain in the operated leg. She scheduled the same on the other leg on her next day off, and a similar operation was done on the other leg a week later. She came back for removal of the staples, and with a great smile on her face said, "I took a long flight with no pain in my legs whatsoever."

FOLLOW UP

A month later she came for a follow up visit and told me that her supervisor almost fell out of her chair when she told her that she could take long flights from then on. Her supervisor asked her what kind of miracle happened so suddenly that she was volunteering for long flights instead of begging to be only on short ones. Diane explained to the supervisor what had happened. Her pain of five years was relieved by a ten minute operation on each leg.

CONCLUSION

The secret of success in Diane's case is what I advise in this book: Look outside the box. The box is the x-rays and lab results. They is no replacement for an accurate history and intuition. She passed on the praise to her fellow workers, and many of them became my patients.

NOTE

In my forty years of practice in Chicago, I never advertised. Most of my patients are referred by others.

ARM COMPRESSION

"Simplify, modify and apply."

—Denton Cooley

THORACIC OUTLET COMPRESSION (TOS)

The symptoms of thoracic outlet syndrome are pain, paresthesia and weakness in the arm, including the fingers. These symptoms characteristically happen on arm elevation, such as reaching for something above the head. TOS occurs more in women than in men, and they can feel the symptoms if they're combing their hair, holding a newspaper or magazine or even driving a car.

The diagnosis is made by asking the patient to elevate his or her arms and exercise them by opening and closing the hands. The affected arm becomes numb and painful within minutes and must be dropped to rest it. In a few cases the pulse disappears upon elevating the arm, leading to the assumption that the cause of this condition is ischemia. The previous description of the pulse disappearing when the patient lifts his arm happens in a few patients. Now the symptoms are known to be pressure on the brachial plexus between the first rib and the clavicle is the real cause of the symptoms, yet the previous assumption that it is caused by interrupting the circulation through the operation on the back of the vascular or thoracic surgeon instead of the neurosurgeons.

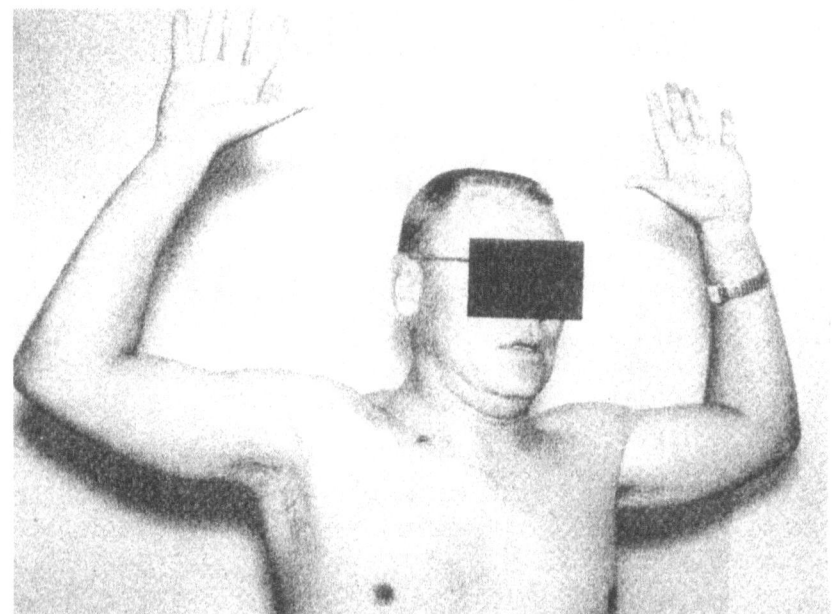

Figure 31: An anonymous patient demonstrates the elevated arm stress test. (Dr. Roos, Vascular Surgery)

The cause of this condition is vague. Previously it was thought it was caused by diabetic neuropathy, cervical disc and arthritis. When none of these causes are found, it's referred to as psychosomatic.

Now we know that the symptoms are most commonly caused by compression of the brachial plexus between the first rib and the clavicle. Previously it was attributed to pressure on the subclavian artery. For this reason the treatment became the domain of the vascular surgeon and not the neurosurgeon. The scalene muscles are anterior, middle and posterior. They connect the neck with the first and second ribs. Their contraction elevates the first rib close to the clavicle and narrows the space between them, causing pressure on the brachial plexus.

The symptoms of TOS are usually preceded by a history of some kind of trauma to the clavicle, chest wall and sometimes to aggressive sports such as batting the ball repeatedly during baseball.

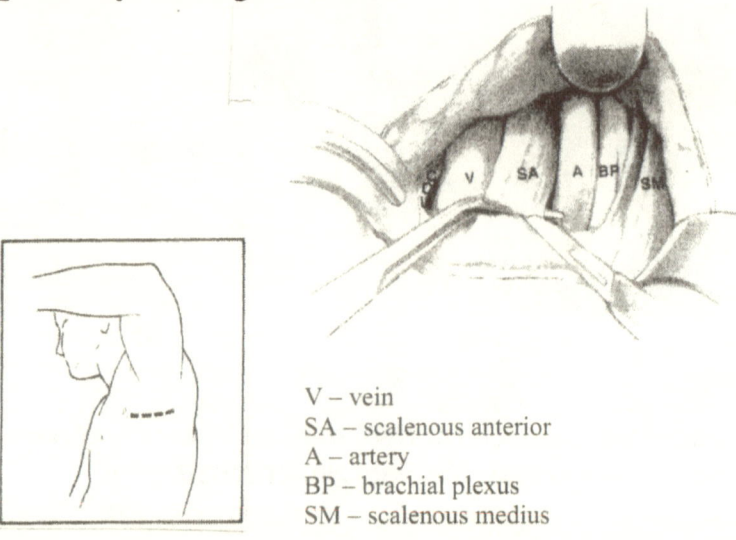

V – vein
SA – scalenous anterior
A – artery
BP – brachial plexus
SM – scalenous medius

Figure 32: The original operation started from the upper border of the first rib and cutting the scalene muscles with a scalpel. A slip of the scalpel which hits the artery or vein would cause uncontrollable bleeding.

THE TREATMENT

To relieve the pressure caused by narrowing of the space between the clavicle and the first rib, one of them should be removed.

Removal of the clavicle is rather unsightly. Removal of the first rib used to be approached either from above the clavicle through the neck (supraclavicular) or below the clavicle through the anterior chest wall (infraclavicular) or through the posterior wall of the chest (the posterior approach). This involves cutting the muscles in these locations to get to the first rib.

Through the creativity of Dr. David Roos of Denver, Colorado, the axillary approach was discovered by him, whereby the incision is done below the axillary hair line hidden under the armpit. It's cosmetically more acceptable. In this approach the surgeon widens the space by retracting the anterior muscle of the chest (pectoralis) and the posterior muscle of the chest (latissimus) and then works his way up to the first rib. When he finally finds it, he has to remove its attachments, mainly the anterior, middle scalene and subclavian muscles. In Dr. Roos' approach, this dissection starts from the superior edge of the first rib.

The operation done this way is not as simple as it seems because any slip of the scissors or the scalpel, while freeing the upper edge, could injure the subclavian artery causing severe hemorrhage, or the brachial plexus causing paralysis. In addition it's difficult to teach, because the surgeon is working through a narrow space and the assistant is pulling from the side and cannot see what the surgeon is doing. This used to intimidate many surgeons and made the operation practiced by only the few of us who had the courage to do it as described by Dr. Roos and emphasized by the diagrams. Thus, few patients could have access to someone to do this needed treatment.

SALVADOR'S MODIFICATION OF THE ROOS PROCEDURE

Inspired by Dr. Cooley's dictum to **simplify, modify and apply**, I thought of a better way to make this operation safer, and popularize it so that more surgeons could feel safe doing it and more patients could benefit from it.

In my modification, when I reach the first rib, I don't start from above, where all the vital structures are located, but rather start from the lower edge, cutting through the muscles between the first and second rib where there are no blood vessels nor nerves. Then I pass my fingers to the under surface of the first rib to free it from adhesions. Next, I pass the index and middle fingers between the scalene muscles to protect the subclavian artery and brachial plexus. Then, detaching the scalene muscles from the first rib can be done without fear of injuring the underlying vital structures. Once the first rib is detached from its muscles and freed, it can be safely cut from its sternal and vertebral ends and removed.

I presented my modification in various conferences, from South America to Japan, and to Europe and the United States. Eventually, it got published in the Annals of Thoracic Surgery during a discussion on the subject.

Many doctors in the United States and abroad asked me to demonstrate my modification technique to them, which I did, and that gave them confidence to try it themselves. The following patient story will serve as an example.

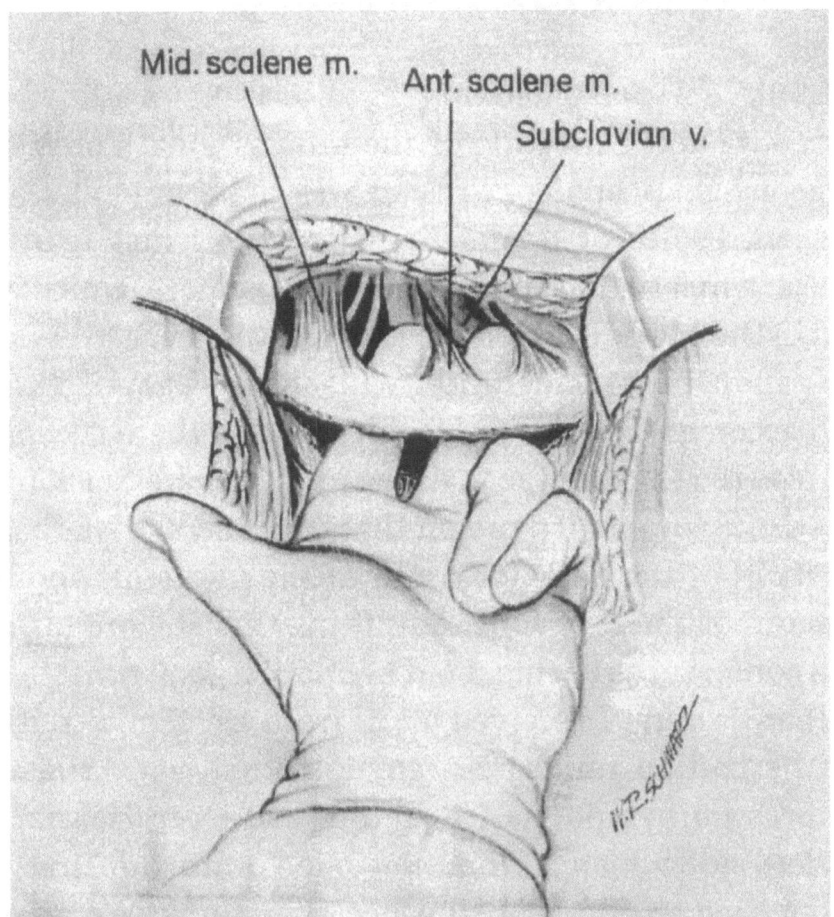

Figure 33: Finger dissection proceeds to push vein, artery and plexus from superior border of first rib. Then pass your two fingers to separate the vital structures bluntly and cut the anterior scalene safely between your two fingers.

INTRODUCTION TO SEHAM'S STORY

In 1984 I presented my modification of the Roos procedure at the meeting of the International College of Surgeons in Acapulco. Present was a prominent Cairo surgeon, Professor Raafat Kamal, whom I met again in 1992, at a meeting of the International College, which was held in the Meridian Hotel in Cairo. I was there to submit two papers for consideration, which to my surprise, were both accepted. It was a joy for me to have them accepted, and also to renew my acquaintance with Professor Kamal. My wife accompanied me on that trip, and she was eager to get out of the hotel for a tour of the pyramids and the museum of the Pharaohs. Admitting that she might find such conferences boring, I bid her have a good time.

That evening I received a call from Dr. Kamal stating that he had an important female patient with bilateral compression symptoms. He said she had a "cervical rib" on the left side which he had detached from the first rib without success in relieving her symptoms. We agreed that he would bring her in the morning to my room and if the diagnosis is confirmed, he wanted me to demonstrate my modified technique on her.

Seham was a beautiful lady in her late 40's who could not keep her arms elevated more than a few seconds. We both found her to be a candidate for First Rib Resection. The detached cervical rib was protruding under the skin and annoying her and disfiguring her neck. That evening her husband drove her to Dar el Shifa Hospital where professors moonlight. Dr. Kamal ordered the x-ray, EKG

244

and blood tests. The next morning her husband drove me in a Mercedes Benz, to the hospital. We reviewed the test results and agreed to operate on her the next day, whereby I would demonstrate my simplified technique to Dr. Kamal. Using my simplified technique would usually take me about half an hour to do the operation.

THE OPERATION

We placed her in a semi-lateral position with the left side up. A resident assisted us in elevating the arm when needed to open up the space between the first and second ribs. I cut through the intercostal muscle between the two ribs, and there was practically no bleeding. Through that space I passed my index and middle fingers under the posterior surface of the first rib, freed it from its pleural adhesions and brought them between the anterior scalene muscle. Protected by my fingers, the scalene muscle was cut without injury to the subclavian artery. I similarly cut the subclavius muscle covering the internal jugular vein and the middle scalene muscle covering the brachial plexus.

To add safety I placed gauze under the two ends of the first rib and cut it at its sternal and vertebral ends, removing it completely. Then I proceeded to remove the cervical rib. The operation was successfully performed in about forty minutes, because the headlight was electric instead of being fiber optic, so it was dim. Also the instruments I used were different from my usual ones. Professor Kamal took some dressings from the Mayo stand and applied a pressure dressing, and we told the resident to watch her and call me if there were any problems.

There were no calls or complications and the next morning her husband drove me again in his luxury car through the heavy Cairo traffic to the hospital. The pressure dressing was removed and I asked to replace them with new ones. To my surprise, the nurse told me that she had none, adding that every surgeon in Cairo brings his dressings with him! The patient's husband then ran to the neighboring pharmacy and bought me the needed dressings. On the seventh day we removed the staples and Seham was able to use her arms without restriction, pain or numbness.

A SURPRISING THANK YOU

On the eighth day, we were scheduled to fly from Cairo to Tel Aviv and her husband hosted a party for me and my friends, who had graduated the same year from Cairo University. Before boarding, he gave me a thank you card. My wife insisted on opening it as if she felt there was more than a note inside. She was right. There was a thousand dollars in cash. She took the money and ran to one of the airport jewelry stores, and bought one of those Cleopatra bracelets which go around the wrist like a snake.

During our stay in Tel Aviv, her husband called twice to thank me and assure me that he and Seham were happy about the results of the operation. But the story didn't end here and there were more surprises in store.

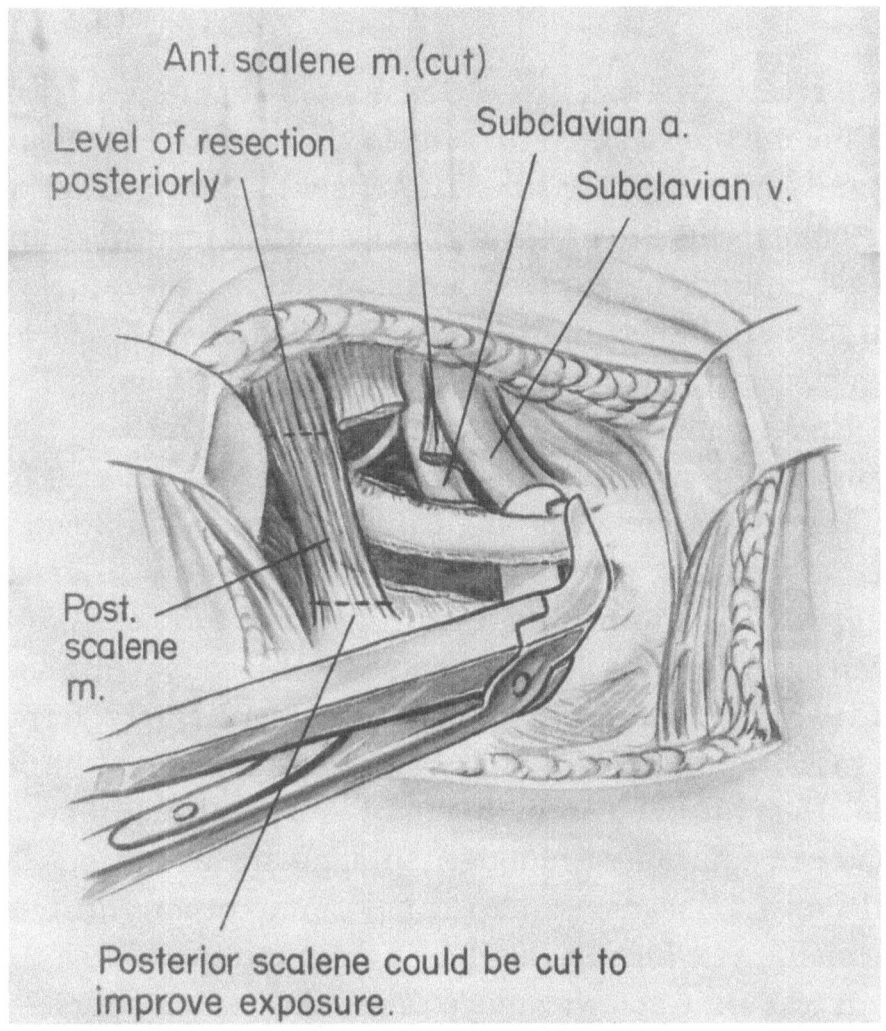

Figure 34: Resection of the first rib anteriorly with periosteum intact.

ONE YEAR LATER

I received a call from Seham's husband in Vienna telling me they wanted to come to Chicago so that I could operate on her other side. They flew to Chicago and we welcomed them to our home. They had no insurance and to get around this problem, I scheduled her for same day surgery at Mary Thompson Hospital, which was the least expensive in town.

Under general anesthesia, the surgery went very well, as usual with my technique, and I took her to our home for recovery. In a few days, the pain disappeared and I took her to visit my sisters who had learned how to speak Egyptian Arabic mostly from Egyptian movies. While chatting with her, they learned she was a cousin of Jehan Sadat. This explained to me why Professor Kamal referred to her as an important lady when he first brought her to me. Her husband was the main hospital supplier of medical equipment, modern diagnostic inventions as MRI's, CT scans, microscopes and other hospital furnishings, which he purchased from Europe.

It was Dr. Cooley's advice to me that, when operating on a celebrity, not to charge them, but to let them offer as payment an appreciation for the benefit resulting from your services. So, when Seham's husband brought up the subject of fees, I insisted that my services were free and I was honored at their coming all the way to America, and choosing me to do her second surgery. Three months later, I received a parcel with fine Viennese crystal and China ware, equaling greater value than I would've charged.

CONCLUSIONS

We're all grateful for Dr. Roos for suggesting the first rib resection could be done through the axilla. He was rewarded for his discovery by being so famous that patients came from everywhere to be operated on by him. But other surgeons were intimidated by the seriousness of any complications that could occur. All that I did was to add a simple modification to his great discovery which made the operation safer, and encouraged other surgeons to do it.

Another contribution I discovered by accident for recurrence of the compression symptoms after a successful first rib resection is different from other suggested procedures, hence freeing the brachial plexus from adhesions, etc.

I was led to believe that the recurrence of symptoms were due to the scalene muscles, which have been detached from the first rib, reattached themselves to the second rib. The second rib was now acting like the first. The simple removal of the second rib was a cure for the recurrences.

SECTION X
WAR CASUALTIES

WAR CASUALTIES

INTRODUCTION

Sometimes things happen in our lives that seem disappointing and not meshing with our best interests. But with patience and time, we eventually look back and realize it all served a good purpose.

As the Bible states, *"Everything is working for the good of those who love God and fit in His plans."*

And remember always, when a door closes, a wider door opens, leading you to greater possibilities.

EYEWITNESS TO HISTORY

My grandfather's parents and their family moved to Taveria, Palestine in 1875. In 1899, the British occupied the Sudan and my grandfather followed, setting up a business in Khartoum and eventually nurturing a family. I was born in the Sudan in 1930 and studied medicine at Alexandria University in Egypt, graduating in 1956. I followed that with a two-year internship in the Sudan and two years in England, finally completing my training in general surgery at Tel Aviv University's Beilinson Hospital in 1966, (It's now called Yitzhak Rabin Hospital).

That same year, Dr. Morris Levy was returning from the states, after studying heart surgery in Minneapolis, and was appointed chairman of a new department of cardiac surgery at Beilinson. Curious to learn about this

253

newly developed art of surgery, I put in an application to work with him and was overjoyed when he accepted me.

Unfortunately, the director of Beilinson Hospital had already promised my place to someone else for political reasons. I was brokenhearted but I could not foresee this would start a chain of events that positively changed the direction of my life. When that door was sadly shut in my face, another door was waiting, leading me in a different path for my life.

The only other place to receive training in cardiac surgery was Tel Hashomer Military Hospital. The Chief of Cardio Thoracic Surgery was Professor Pawzner, who by chance had an opening to replace a Dr. Lieberman who was sent to intern at The Cleveland clinic.

Unlike Beilinson, which was a very modern eight-floor hospital, Tel Hashomer was, at that time, a military hospital wherein each department was housed in a set of barracks. Barrack 10 was designated solely for cardiology and cardiac surgery, and Professor Henry Neufeld was in charge of the cardiology unit. Neufeld wrote one of the first books about congenital heart disease, and I learned a wealth of information from watching him.

At that time I was preparing for the ECFMG (Educational Council for Foreign Medical Graduates) where I needed to train in the states.

The year between June 1966 and June 1967 was by all accounts a politically quiet year. Then, in June of 1967, the Egyptian dictator, President Nasser, closed the exit to the Red Sea which Israel used as its major import/export hub, and declared war on the tiny nation. He managed to convince Lebanon, Jordan, Syria and Iraq, to join him, and it was a difficult three weeks for the people of Israel while negotiations to retract that declaration of war went on in the United Nations. The Israelis were prepping shelters and practicing air raid routines, and Israeli Prime Minister Levy Eshkol seemed lost as to the best way to respond to such aggression. Despair loomed over Israel and Eshkol composed a Unity Government, naming Moshe Dayan as Minister of Defense. The Chief of Staff of the Army at that time was Yitzhak Rabin who organized units in the Sinai front and had them ready to respond to any attack from Egypt. The Air Force chief was Mr. Weisman, the nephew of Israel's first president, Chaim Weisman. The cabinet gave the Israeli Air Force the audacious command to destroy all of Egypt's airports, aircrafts and air arsenals. Over four hours of unceasing bombing left the Egyptian Air Force in tatters, depriving them of the air mobility needed to prevail.

Dr. Cheba came in the following morning and told us, "Don't worry, we won the war."

The hospital, however, now had to deal with the ramifications of that victory. Egyptian soldiers poured in with wounds to the chest, belly, head and every other place imaginable. Because other surgeons were working day and night with injured Israeli soldiers, Dr. Cheba, in his capacity as director of the hospital and chief surgeon general, assigned me to take care of the wounded Egyptian prisoners of war, because I had studied in Egypt and could easily communicate with them. They were air lifted to Tel Hashomer Military Hospital.

When I went to see them, they were in a state of panic, fearing the Israelis were going to massacre them. When I spoke to them in Arabic and told them I had graduated from an Egyptian medical school, they relaxed. I assured them that they would be treated just like the wounded Israelis, and I will treat them as patients with complete disregard to their political status. They embraced me, kissed my hands and without delay, I began the tedious process of planning their treatment.

Along with a group of young doctors, I worked day and night triaging them, operating first on those in danger, followed by the less seriously wounded. By the end of the war, they had all recovered except one who had chronic empyema. Eventually, Israeli and Egyptian soldiers were playing football together. This gave me great experience in trauma treatment.

The victory on the Egyptian front motivated the Israelis to defeat Syria and liberate the Golan Heights and next to free Jerusalem.

A month later the IDF (the Israeli Defense Force) flew a group of sailors suffering from pressure injuries to Tel Hashomer by helicopter. They were cruising across the Egyptian coast when the radar told them a missile was coming in their direction. The captain ordered them to abandon ship and jump in the water. The boat was hit and sunk, but the sailors who survived suffered from water pressure injuries, ranging from ruptured bowels to pneumothorax, as well as other vascular trauma.

THE BIG EVENT

Two months after the war ended, I was again on call at the ER and the tranquility was interrupted by an eruption of noise from cars and horns. I thought it was either a terrorist attack or a bad accident. Instead, a convoy of paramedics and soldiers were escorting the hero of the war, General Moshe Dayan, who was on a stretcher, badly injured.

First thing I thought was it had been an attempt on his life, but the paramedics told me that he was digging in a historical excavation, which was his hobby, and a wall collapsed on him. They said he was buried in the rubble and would have been dead if the children playing in the area had not recognized his eye patch and alerted their parents.

When I greeted him, he responded by winking. I asked for an immediate chest x-ray and started taking his vital signs. He asked for something to ease the pain but I told him this could not be done as his blood pressure was only 80/60. I sent his blood to be typed and cross matched and then inserted an IV. When he stabilized a bit, I ran to inform my boss, who was in the middle of a weekly meeting of the department heads, that Dayan was in our ER and badly injured.

On the way to the ER, I filled him and Dr. Cheba in on Dayan's condition. Some ribs were fractured, as well as his vertebrae; there was also evidence of increased mediastinal and cardiac shadows. When Dr. Cheba saw the x-ray, he asked Dr. Pawzner, "What do you think?"

"I think he ruptured or dissected his aorta." Pawzner replied.

"What do you think we should do?" Cheba asked.

Dr. Pawzner replied, "The most capable doctor to handle such a situation is Dr. Denton Cooley of Houston, Texas."

Dr. Cheba asked Pawzner to call and explain the situation to Dr. Cooley. Over the phone, long distance, he suggested stabilizing Dayan and flying him to Houston where he would be more comfortable with his own team and equipment. If this couldn't be done, Dr. Cooley was willing to fly to Israel.

By this time, Dayan had received 3-4 units of blood, was sedated showing normal blood pressure and seemed stable.

It was bout 10:00 PM when Doctors Cheba and Pawzner asked me to stay and watch the general throughout the night and report any change in his condition.

General Dayan got to know me and asked me who his nurse was. She sat next to him and he chatted with her about women in the army. He wanted to know if she finished her army service, what unit she was in, etc. He asked me to stay in the next room and the nurse would call me if there was a change in his vital signs.

At about 1:00 AM, the radiologist on duty came to see me. I was wide awake in Dr. Pawzner's room, reading and mentally preparing myself to fly with General Dayan to Houston. The radiologist asked me how the chest x-ray was taken and I said lying down. He told me an upright chest x-ray was needed. I told him to call his boss and ask his permission, since Dayan was a special kind of patient and I didn't want to take the responsibility should he collapse.

The conversation between the radiologist and his boss was very brief, as his boss dismissed performing an upright x-ray on a patient with fractured ribs and spine. The radiologist asked me if he could talk to General Dayan personally, and the General agreed. The radiologist suggested the General be transitioned to a movable bed, tied to it, and to turn the whole bed upright without Dayan stepping on his feet. Dayan agreed to this and the radiologist was able to take a few shots of the general's chest with the bed three-quarters upright.

It was amazing to see the results of the "controversial" x-rays. His heart and aortic silhouettes looked normal, and all the blood had accumulated at the bottom of the chest. It was clear the bleeding was coming from fractured bones and had nothing to do with the heart or the aorta.

When we gave General Dayan the good news, he flashed us one of his charming smiles. By that time, he was ready to sleep and we decided to give the good news to our bosses when they arrived in the morning.

I told Dr. Pawzner first, who relayed the finding to Dr. Cheba, and they both called the chief of radiology to congratulate the young radiologist on his determination. A trip to Houston would not be necessary after all.

General Dayan continued to improve. The bleeding stopped, and he asked how long it would take for his fractures to heal. It would be three weeks and I was asked if I could be in charge of his care during that time, to which I graciously agreed.

Many people, of course, wanted to see him. The first one he agreed to see was his daughter Yael. Then he saw Shimon Perez, who was followed by the Mukhtar, the local

Moslem authority. And Prime Minister David Ben Gurion visited him on the fourth day.

Ben Gurion had a private twenty minute visit with the General. Dr. Cheba joined Dr. Pawzner, and both men knew the hospital staff would like to have a personal encounter with the great Ben Gurion. Dr. Pawzner introduced us and he shook hands with everyone.

One of the high points of my life was to meet this man with the penetrating look who commanded such awe and respect. Dr. Pawzner told him I was the doctor who admitted General Dayan into the ER and that I was born in the Sudan. I was so flattered when Mr. Ben Gurion kept holding my hand, and asked me about the Jews of the Sudan.

It was amazing to meet the man who dreamed of a modern day Jewish state, and whose dream was eventually realized. He wanted it to be a melting pot for Jews, dispersed all over the world.

After three weeks, when we went to see Dayan on the usual morning visits, we found the General dressed in his military uniform and ready to leave. Dr. Cheba told him he hadn't been discharged yet.

"When I came here, you told me that my fractures would take three weeks to heal and today is the end of three weeks," Dayan said.

Once he left, things went back to normal, and a quiet atmosphere again prevailed over the hospital.

It was 1967 when I passed the ECFMG exams, and Dr. Cooley, a pioneer in heart surgery, started a series of heart transplantations in every terminal heart patient sent to him. When one transplant was rejected he was the first in the world to replace it with an artificial heart, pending a donor. His name was all over the newspapers and TV. I felt a compulsion within me to work for this great surgeon and to get to know him and the secrets of his success. I wrote to Cooley asking if there was a possibility that I study under him after Dr. Lieberman returned from Cleveland. I reminded him I was one of the doctors who were going to accompany Moshe Dayan to Houston.

A great surprise came three weeks later when I received a personal letter from him informing me he had an opening of a one-year window between June 1968 and July 1969.

I ran to Dr. Pawzner euphoric. He sat me down, said, "First of all, wait until Dr. Lieberman returns from Cleveland and work with him for a few months to learn what he learned."

The second condition was that he wanted me to spend a year with Dr. Starr before going to Cooley. Dr. Starr was the first to invent a plastic valve to replace a damaged heart valve. Pawzner had gone to Portland, Oregon in the late fifties and studied valve replacement and open heart surgery under him. Dr. Favaloro, an Argentinean doctor visiting in Cleveland Clinic got interested in coronary artery disease and was one of the founders of coronary bypass surgery. Dr. Lieberman was sent to learn this technique in the Cleveland Clinic.

Dr. Lieberman returned and I learned a lot from him about bypass surgery. In March of 1968, I took my family and flew to Chicago to spend a few days with my sick father, mother and sisters.

I had a special ticket called **Visit America** which allowed me to visit eight cities in the United States. I used this to fly to the Mayo Clinic and spend time with Dr. Magoon, a gentleman and great pediatric heart surgeon. Then I flew to New Orleans and visited the Ochsner Clinic. After that I went to Boston and visited Massachusetts General, Beth Israel Hospital, and others. Next, I went to New York and visited the heart unit at New York University Hospital in Manhattan and Dr. Kantrovich of Brooklyn Hospital who, during a cardiology meeting in Tel Aviv tried to recruit me for research on cardiac transplantation in dogs and freezing the hearts pending re-implantation in another dog. I apologized and told him I was already committed to work for Dr. Cooley in Texas. And in April of '68 I arrived in Houston. All this would not have happened if my appointment in Beilinson Hospital had not been cancelled.

THE CONCLUSION

Once again, this proves that *"When one door closes, another door opens that could lead to a better destination than the one that was closed."*

PART II

is of great interest for both patients and doctors

FOR THE PATIENTS

to choose a doctor who can cure the incurable and stand beside them in such situations.

FOR THE DOCTORS

to achieve miraculous cures by awakening those talents inside them and use them.

INTRODUCTION

> *"Medicine is an art, not a trade,*
> *a calling, not a business,*
> *a challenge in which your heart*
> *will be needed equally as your head."*

—Sir William Osler

Medicine is a great intellectual science, and intellect is a quality of the mind. To achieve maximum healing, knowledge should be complimented by specific traits, some of which are inherent in our souls. In this section, we'll try to discover them, and find out how they relate to the patients' stories.

Most medical schools teach their students the science of medicine but they do not teach them *how to think beyond the curriculum or how to interact with their patients*. I hope this book will fill this gap.

To bring attention to these qualities needed to cure the incurable, we will analyze each quality in detail in this section. To achieve miraculous cures, we must develop these traits in ourselves first. We'll emphasize the importance of each of them so that you can apply them when needed. In addition to that, you'll find information regarding the doctor-patient relationship, the role of medicine, the use of complimentary methods, and we'll explore the value of spiritual power. We'll also talk about the universal mind, the interconnection between our minds, as in ESP, and its use in calling for help. There are sections on the importance of intuition, in making medical decisions, and how it can inspire us.

Creativity is the essence of discovering new ideas and discarding old ones that do not benefit our patient. So, let's go discover these traits. And the first traits needed to achieve miraculous cures are love and the compassion to be creative.

For reference, all of the traits are listed below with their corresponding page numbers:

LOVE & COMPASSION

"When a patient is brought in the operating room, as soon as I touch his heart, he touches my heart, and I give him my best."

—Denton Cooley

Love is a common thread without which no healing is complete. Love breeds compassion, intentionality, devotion, diligence, and resolve.

Compassion is the soul of medicine, and without it medicine is a dead science. Intentionality means determination to find a cure where there is none in medical literature.

An example of the value of perseverance was an extra two hour search for a blood vessel to carry the circulation in Bernice Herring. Neither the anesthetist, my assistant vascular surgeon, nor the radiologist who prepared her for amputation, could deter me from pursuing my search. The result? Saving her leg for the rest of her life.

Another example is the love for my aunt who was scheduled for amputation above the knee. I was inspired to find a successful way of amputating only her gangrenous toe without risking the gangrene spreading to the thigh, as anticipated by specialists in the field.

Another example of the value of a combination of compassion and perseverance was needed to save lives of patients in coma when everyone else declared them hopeless.

Love is a healing energy, and miracles are expressions of love. Love respects the uniqueness of every individual patient and empowers them to take responsibility for their own well-being. Such unconditional love can be transferred from healer to patient to create trust, which is of utmost importance in the doctor-patient relationship. A doctor's love for his patients helps them love themselves, and is a strong motivation for self-cure. It creates a reciprocal state of harmony and trust which enhances compliance and facilitates healing.

Love is selfless, and those who truly love graciously give of themselves to others. Of all the traits that heal, awareness of Divine Love is the most effective.

Compassion means a genuine concern for the pain of others. Without compassion, the medical practitioner lacks the will to seek an answer, to research the problem or to find another way of salvaging a patient who has been deemed beyond hope. A compassionate response awakens the soul of a seriously diagnosed patient. It triggers a timeless connection between the souls. A simple hug can assure the patient he is not working alone towards recovery.

Empathy comes with compassion. It means feeling the pain and suffering of someone and trying to extinguish it. Empathy is complete understanding between two human beings. It's a non-verbal interchange of feelings, beliefs, and attitudes between doctor & patient. It creates a subtle communication, significant in their relationship.

The best way to show a patient that you care is to **listen attentively**. This leads to better understanding and communication, and can be psychotherapeutic and reassuring. It helps the doctor focus on the patient's story in a way that could lead to a correct diagnosis.

Some doctors are taught not to get emotionally involved with their patients. This approach is an attempt to bring a dispassionate and scientific healing of illness, without attention to his emotions or spirit. But in the last decade research has shown that emotional factors on the part of the health practitioner affirm the effect of the treatment, reduce recovery time and medical costs. This brings about trust, which helps the patient to comply better.

When medicine fails to find a remedy to cure a patient, it is always important to display love. You must pray, and ask for guidance. This will inspire you toward a different method that'll positively affect the prognosis.

Achieving miraculous cures demands not only scientific skill but an open heart and mind.

"And though I may have the gift of prophecy and understand all mysteries and acquire all knowledge, if I have no love, I am nothing."
—I Corinthians 13:2

"Abide in faith, hope and love, but the greatest of these is love. Love never fails."
—I Corinthians 13:13

"If you listen to the patient carefully, he'll give you the diagnosis."
—Sir William Osler

The Power of Love

There is no difficulty that enough love will not conquer.

No disease that enough love will not heal.

No door that enough love will not open.

No gulf that enough love will not bridge.

It makes no difference how deeply seated may be the trouble, how hopeless the outlook, a sufficient realization of love will dissolve it all.

If only you could love enough you could be the happiest and most powerful being in the world.

—Emmett Fox

GOOD DOCTOR-PATIENT RELATIONSHIP

"Medicine is not only a science, but also the art of letting our own individuality interact with the individuality of the patient."
—Albert Schweitzer

"The physician should not treat the disease but the patient who is suffering from it."
—Moses Maimonides

The doctor-patient relationship should be based on **love**, **respect** and **trust**. A patient's trust is very important if the patient is expected to follow his doctor's opinions, especially when they differ from others. When a patient goes to a doctor, he goes to him because of his unique personality, knowledge, experience and reputation. His trust is important in accepting the doctor's information, opinion and possible modifications. For example, many patients came to me with breast cancer for lumpectomy only when others recommended mastectomy.

There were countless times when I challenged the benefit of certain customary procedures referred to as "protocols." I doubted these established principles and proved that lumpectomy is better than mastectomy. I think that the current post-operative protocol does more harm than good, and I don't recommend it. I also rejected the repeated notion in surgical literature that "lymphoma is not a surgical target," and discovered a method to cure it safely in a surgical manner as well. The same holds with metastatic cancer, once considered one step away from the grave. Protocol suggests post-operative irradiation and chemo therapy as routine after every lumpectomy for breast cancer. Radiation by itself is known to produce cancer, and chemo has its toxic effects. And both modalities can diminish the resistance of the patient, leading to recurrences of cancer, rather than preventing it. They can also predispose them to infections such as pneumonia and other respiratory complications. They should be reserved for metastatic cancer, but if they're used prophylactically, they can lose their effectiveness when they're needed.

I had to explain my opinion to my patients and give them the choice to follow the protocol, or refuse it and get back to their normal lives. Those who trusted me and followed my opinion had better results than those who followed the protocol.

I'm the type of doctor who just cannot accept as fact everything that has been published, and if I doubt its effectiveness, I relate that to my patients.

Many medical schools like Harvard, Johns Hopkins, and Duke, teach their medical students early to think innovatively, rather than follow the texts blindly. These schools instill in them the notion that medicine is a progressively changing art, not a static science. It's an art that combines knowledge with intuition, judgment, open-mindedness and creativity.

But in order for the patient to accept his doctor's personal advice and unique methods, he needs to trust and respect his doctor, and know that he feels love and compassion for him. The doctor should humbly explain to the patient his reasons, and give him an informed choice, and document it. This is especially important in this era of litigation, in which trial lawyers have imposed certain "standards of care" on the medical community, depriving our patients of more advanced health care and hope.

One the flip side, some doctors regard optimistic views as false hopes, and portray the worst case scenario with the rarest possible complications. They do so in the name of "truth and informed consent." In my view, this is tactless and destructive.

It is now known that hope breeds faith as well as the expectation of a good outcome. It's also known that such feelings help produce cheerful chemicals in the brain, like endorphins and dopamine, which positively affect the body. The destructive influences of fear, worry, and anxiety can be reversed with *the relaxation response*, filled with the benefits of prayer, meditation and support groups.

When patients are suffering from pain and debility, they look to doctors and nurses to reinforce their belief in medicine's power, and solidify their expectation that they will benefit from the intervention.

Informed consent, which lawyers invented, emphasizes the few negative complications that may occur during surgery and ignores the 95+% of success. In my view, this is misinformed consent. In this day and age, with emphasis on mind-body effect, destructive information may convince the patient that the worst will happen. Sadly, the body responds to that. Hope is important in quelling those patients' fears.

None of us wants to give any patient false hope, but we also do not want to instill unwarranted fears. False hope is when the doctor hides the truth from the patient. True hope is when the doctor informs his patient of his condition and gives him the choices available and the support he needs.

A good doctor-patient relationship is perhaps the physician's most powerful tool for dispelling panic, and mobilizing the patient's body to exert maximum force against disease. This partnership might be the strongest factor in the healing equation.

This was illustrated in the story of the Mexican Cowboy.

In trying to obtain an "informed consent," an anesthesiologist tried to scare him into thinking the operation was dangerous and that he could die or remain on a respirator for the rest of his life or worse. But, no matter how much he tried to dissuade him from going to surgery, that cowboy's response was, "I trust my doctor."

As a result of this emphatic trust, along with the patient's positive attitude, as mentioned before, the surgery went smoothly, deflating the cyst and placating it. His suppressed lung expanded, and the patient made a spectacular recovery. He left the hospital a week after surgery, without which he would've died from respiratory failure.

At the present time, doctor-patient relationships are suffering terribly. With new diagnostic tools, like CT Scans and MRI's patients complain that medicine has become mechanized and lost its personal touch. Moreover, the results of these tests are not always accurate, and could be misleading. Patients are complaining that doctors don't listen to them, or even examine them, thus resulting in a lack of confidence in not just the doctor but the healthcare industry as a whole. To avoid the consequence of a false diagnosis, an accurate history should be obtained by listening carefully to the patient and comparing their complaints to the test readings. If the test reports clash with the patient's complaints and/or condition, we should look outside the box to avoid being misled into a false diagnosis.

Attentive listening is a sign of love and respect, and the greatest compliment we can pay a patient.

Dr. Jerome Groopman is the Chairman of Medicine at Harvard Medical School. In his book, *How Doctors Think*, he addresses this problem. He says misdiagnosis can result from a lack of detailed history and sole dependence on diagnostic modalities. Dr. Groopman agrees that a doctor should rely on his intuition, logic and common sense to solve the discrepancy.

Most doctors stereotype patients with the most common medical conditions rather than looking for exceptions. They are taught evidence proves the diagnosis, even when the evidence may be wrong! Moreover, this outdated philosophy starts with the conclusion, which is the diagnosis, and works backwards from there, seeking signs that support that supposition. The doctor must consider all the possible data before arriving at his conclusion. We have to keep this in the back of our minds.

An example was Diane, the airline stewardess who complained of pain and numbness in her legs during long flights. Consultants found that all the tests, including angiograms, were negative. They decided that nothing was wrong with her and completely missed the diagnosis of *compartment compression*.

Another example of this was found in the story of Lida, whereby the CT scan and MRI of the brain were read as Multiple Sclerosis. She became so convinced that she had the disease that she actually acquired its symptoms. A proper history revealed that this diagnosis was incorrect and she actually had chronic fatigue syndrome. She was treated accordingly and it led to a permanent recovery.

A poor doctor-patient relationship can also result in mistrust. This dissatisfaction can lead patients to seek alternative medicines, produced by some natural medicine companies that sell patients untested or ineffective medication. The use of alternative medicines should be based on proof of their effectiveness, by research and at least animal trials.

In this Information Age, patients should work in partnership with their doctor. They should equally love, trust and respect him in order to achieve optimal health. They can be equally knowledgeable of their choices, and doctors should humbly share their information with them, especially when asked about alternative medicines. As mentioned before, I benefited from patients' questions about alternative medicines, and instead of dismissing their inquiries, I researched them. Then I came back with an answer regarding its true value. Some of them proved to be of worth, and I included them in my practice, e.g. chelation therapy and antioxidants. These alternatives are now referred to as *complimentary medicine*.

INTUITION

"When you are inspired...dormant forces, faculties, and talents become alive, and, you discover yourself to be a greater person by far than you ever dreamed yourself to be."

—Pantanjali

Intuition is a process of reaching accurate conclusions from inadequate information. It tunes us in to the currents of universal energy, helping us to tap into a power greater than ourselves in order to create and manifest something positive. It's possible for us to ask questions and get answers by accessing that universal energy through our intuition.

Intuition is also essential to patients and doctors in obtaining spectacular cures. Patients should be aware of subtle changes in their body functions and seek therapy for them. Often, wives are more sensitive to their husbands' physical condition and may initiate the necessity of medical care. They sometimes could be lifesaving, as during an acute heart attack when the husband is in denial. In such situations a man might dismiss symptoms of retrosternal pain as either gas or indigestion, while the wife's intuition could tell her otherwise and take her husband to the hospital to be saved.

For doctors, intuition is important for coming to a correct diagnosis in patients with conflicting symptoms, or those suffering from unusual conditions that are not commonly considered. In such conditions, deep thinking and determination often lead to a hunch that clarifies the situation and guides the doctor to the proper diagnosis and treatments. Such a circumstance occurred to the stewardess with leg pain, and the lady with esophageal spasm. More examples can be found in the previous section of patients' stories.

Examples of intuition are when we make a snap diagnosis as soon as we see the patient, or ever hear from him.

I recall an occasion when a patient called me late one night regarding rectal bleeding. For some unknown reason, I told my wife, an RN, that he might have a rectal tumor. She was astonished at my hypothesis. It wasn't really a conclusion based on facts. It just flashed in my mind that the patient wouldn't call for a usual rectal bleeding that late at night. There must be more to it, and it probably scared him. Surprisingly, after working him up, I found that my snap diagnosis was correct: he had a low rectal carcinoma, to which I had to perform an abdominoperineal resection of his colon a few days later.

Intuition can also be defined as a segment of thinking that cannot be explained. It's the ability to reach sound conclusions from minimal evidence and emphasis. The speed and reliability of finding the right connection cannot be found by analytic methods. It refers to a remarkable mental performance that moves rapidly, yet unaccountably, toward a correct conclusion.

Surgical intuition is a characteristic of mature surgeons with extensive knowledge and experience. Intuitive surgeons can see into the belly simply with a brief history and a careful abdominal examination, without laboratory or x-ray data. With astounding accuracy they somehow know what's happening.

To become a great surgeon, one needs to gain the three features of surgical intuition:

1. Exposure to adequate surgical volume,

2. Surgical curiosity, including avid reading of medical journals, open discussions with consultants, and continual growth of one's knowledge base, and

3. The ability to look back on cases and review them objectively, particularly with regards to complications.

Over the years, the analytical side of medicine has been emphasized with great strides and innovations, but the intuitive side which encourages judgment, experience, and common sense, has largely been ignored. Recently however, the trend has been reversed in favor of intuitive diagnoses. The use of intuition is not limited to medicine. We use intuition when we are attracted to some people and reject others, most of the time justifiably without knowing why.

Dormant intuition can also be developed when we open our souls and tune into the infinite intelligence of the universe, and develop the ability to communicate with each other through the use of telepathy and ESP. By the same token, it's my experience that whenever you're in need of an answer, life will provide it for you through past experience. It has happened to me on many occasions.

For example, when faced with overwhelming massive bleeding, I ask myself how Dr. Cooley would handle such a grave situation. The first thing that comes to mind is to keep my composure and serenity, stay cool, and do my best. After that I can imagine Dr. Cooley performing the operation through me, because I follow his methodology.

Intuitive medical diagnosis is not limited to the medical practitioner, but amazingly, it comes as a gift to some psychics. Edgar Cayce, born in 1877, never studied medicine. Yet, when given the name of a certain patient with certain symptoms, he could go into a hypnotic trance and diagnose the condition with its proper medical name and description. He then sent his extrasensory perception (ESP) to search for a doctor in the states with the most experience in treating that patient's illness. He even went so far as to refer patients to the doctor with the correct diagnosis. This was verified and published in the press. In her book on intuition, Dr. Mona Lisa Schultz suggests there is a psychic in all of us, and intuition works by tapping into the energy of the universe and its wisdom. To others it's a gut feeling from previous knowledge or experience within us, or by telepathic communication with others. There's a place inside of us where we feel our oneness with all, our connection with each other and the universe.

Surgical intuition is the most important quality for the surgeon and the only venue to obtain solutions when there is none other. It is not only a tool for quick diagnosis, but a way to find new surgical techniques were there is no other recourse. In this case, a surgeon can tune into universal intelligence, and the answer will come to his mind.

Ideas and answers could also come through meditation. There are so many ways to meditate, but I prefer to simply go to any quiet place and relax. I become conscious of my breathing, and empty my mind, leaving a space for new ideas to enter. Before long, a solution usually appears. At this moment, you can enter the fraternity of those who do unusual or miraculous cures. It's that simple.

For the doctor, intuition opens the door to creativity whereby he can find new treatments, even when the books say there are none. For the patient, prayer, meditation and use of intuition can guide him to the most suitable surgeon or institution to treat his condition. No patient should surrender his body to a doctor or surgeon with whom he does not feel comfortable or trusting. In this respect, many wives have saved their husbands with immediate care, even when his condition was serious and the husband in denial.

Throughout the stories of my patients you'll find that intuition has helped me greatly in reaching correct diagnoses, designing surgical procedures, and avoiding some routine protocols mentioned in the books which I felt were more harmful than useful.

CREATIVITY
THE ART OF INNOVATION

"Every treatment is given by a loving person directed by God's mind for a creative purpose."

—Charles Baker

Creativity is the energy that leads to new ideas and discoveries. This is done through a process of rethinking, emptying our minds for a while, and then filling that void with a new imaginative picture that evolves into a new technique or structure.

The creative individual is a person who regularly solves problems and finds solutions in a certain domain in a way that is considered totally novel, but ultimately becomes accepted in a particular cultural setting, e.g., medicine. The creative person sees what no one else is seeing and frequently thinks what no one else is thinking.

Everything in this world started from an idea which took form in the mind followed by structure: "The word becomes flesh" or "The formless takes form."

We are in an era of accelerated discoveries, and the pace of life is increasing so much that sometimes it can be difficult to cope with the changes and absorb new information. Through nuclear physics, we are able to multiply organisms through cloning in the hope that it will eventually lead to replicating animals or even humans. This means that we can continue to live through

283

clones of ourselves. But this awesome scientific power should be balanced by spiritual values. There is great potential in the undiscovered part of the universe e.g. the part of the universe that is unseen, sometimes referred to as the *spiritual universe*.

During my fifty-year medical career, I got to know and admire some of the great medical inventors. Dr. DeBakey invented the roller pump that was used in blood transfusions and is now being used in the heart-lung machines for open-heart surgery. Dr. Cooley also pioneered new techniques from the inception of open heart surgery, like aortic cross clamping for cardiac arrest, hemodilution when no blood bank could keep up with his operative volume. He was also the first to use a mechanical heart to bridge the gap between a rejected transplant and finding a more suitable one. He proved, for the first time ever, that life can be sustained by a mechanical heart. Such great discoveries were made by men whose thinking ran ahead of its time. A creative person is one who has insight, who can see things no one has seen before—this is creativity.

Creative people don't live their lives as passive observers, but take a pro-active role in the flow of life and are privileged by a gift from God to be co-creators with Him for the benefit of mankind. They're usually skeptical about what they've been taught and they're thrilled to discover new facts by themselves. Inventors are free from conventional restraints, criticism, and censorship. Their minds are set by a collection of relevant information and persistent attempts to correct a problem with new ideas and solutions.

From the stories you've read, you can see that I put my power of imagination to the test when it was needed. I can now remember the poor nurse who was shot by a BB gun after her evening shift, the traumatic arterio-venous fistula which was deemed inaccessible or metastatic cancers and lymphomas that were thought to be outside a surgeon's domain. I discovered new techniques that resulted in the patients living long, fulfilling lives after being thought to be incurable. Such experiences have made my life worthwhile. When a case is rare and no description is found in literature, it's up to the surgeon's imagination to find a new way to solve the problem. This is how the incurable becomes curable.

When resection of the first rib in the treatment of the Thoracic Outlet Compression intimidated many surgeons away from it, I invented a technique that simplified the operation and made it safer, encouraging other surgeons to go through with it. This made it available to a greater number of patients who suffered from this compression.

Creating a new device or technique usually follows the need for it. As the proverb goes, "Necessity is the mother of invention." It's a privilege and an honor that I was given the chance to use this gift, and I recommend it to all my colleagues. It takes courage to overcome the inhibitions and fears that result from the deviation from the norm. This is the only way medicine can evolve.

A lot of books are written trying to explain the factors that make a person creative. My creativity stems from love for my patients and a determination to help them, as well as the humility to ask for inspiration, which has typically led me to discover new ways to solve their problems.

Some tenets of a creative thinker are: produces many ideas; judges which ideas work and which don't; looks at things differently than others; challenges everything; improves on existing ideas, rather than knock them; burns with desire to be creative; isn't deterred by problems; makes meaning out of unrelated words; refines ideas to perfection. These may help you to evaluate your degree of creativity. Without new ideas the world cannot progress.

SERENDIPITY

"Thought allied fearlessly to a purpose becomes a creative force."
- James Allen

Serendipity is a process by which something is discovered accidentally while looking for something else.

Scientists who made serendipitous discoveries declare that although they look as if they were discovered by chance, they had been preparing their minds towards such incidents. Louis Pasteur, who discovered bacteria while looking for the causative factor in souring wine, said, "In the field of observation, chance favors only the prepared mind."

Sir Alexander Fleming discovered penicillin when his Petri dish of bacteria had mold on it. Fleming said, "If it was not for my previous experience as a bacteriologist, I would have thrown the plate away and would have lost the discovery of Penicillin."

At the turn of the 20th century, a medical student, studying the digestive effects of the pancreatic juice, accidently discovered insulin. There was no treatment for diabetics at that time.

Another serendipitous discovery is that of cisplatinum, one of the most potent chemo therapeutic agents for cancer. As the story goes, while studying the behavior of bacteria in electric fields, researchers found that the platinum found at the ends of electrodes was a potent bacteriostatic agent. This discovery could have been ignored, but since cancer spreads by rapid division of cancer cells, Dr. Lawrence Einhorn of Indiana University had a hunch that platinum might be useful in arresting rapid cancer-cell multiplication. A noted survivor who was saved from testicular cancer with cisplatinum wrote a book about it. He is none other than seven time Tour de France winning cyclist Lance Armstrong.

Serendipity is not confined to scientific discoveries alone. In fact, the word serendipity—the faculty of finding valuable and or agreeable things not sought for—was taken from a Persian fairy tale that took place in an area of Ceylon called Serendip. That story is about a prince who went searching for treasure and always managed to stumble onto greater and more rewarding discoveries that what he sought.

The word serendipity when divided in two, *serene* and *dipity*, means the prince dipped into life, expecting to find something valuable in his lifetime, but came across unexpected discoveries that brought him greater happiness.

Serendipity is now a way of life. People, who adopt it, while doing a certain task, expect something serendipitous to happen. This mindset gives them the advantage of catching, grabbing onto and seizing chance opportunities. Spiritual folks define it as a God-given surprise.

For years I wanted to find a way to decrease the suffering of my patients who died in severe agony following the traditional medical treatment for metastatic cancer and lymphoma. And when the occasion came, it was by accident that I discovered the technique of enucleating the cancer pulps from their capsules, thus making the inoperable, operable, resulting in chemotherapy becoming much more effective because its target became infinitely smaller. All this came to me while opening the capsule in an attempt to take a biopsy. Comparing the few patients I treated with this method to the suffering of others who received the conventional treatment, I know I was guided by the energy of love, and discovered this technique by serendipity.

While happily living your lives, always expect something pleasant—maybe even a miracle to occur. Keep your eyes open. It happens when you least think about it. In short, serendipity is discovering what you are not looking for.

THE VALUE OF POSITIVE AND HOPEFUL ATTITUDES

A positive, hopeful attitude is very important in both patient and doctor. If a doctor lacks hope and confidence in curing a patient, he should send him to a more optimistic physician. Only physicians with faith and confidence in their abilities to heal, and hope in finding a cure, can achieve remarkable healings. But if a patient thinks he cannot be cured, then he cannot be helped.

An attitude of optimism and hope in both patient and physician can produce miracles. In my experience, surgery goes more smoothly, and recovery is more prompt in patients who go to surgery with positive expectations. Their body responds to their positive thinking and acts accordingly. A perfect example of a positive outlook that led to an unusual longevity is that of my cousin Mayer, as told in the section on patients' stories. To keep a positive attitude, a patient needs a supportive group of people or individuals, which in his case is his devoted wife.

The worst negative emotions are fear and doubt. I think that fear of cancer kills more patients than the cancer itself. We have to be very careful and sympathetic when relating a finding of malignancy to a patient and decline from quoting him statistics regarding his lifespan, because he might come to your office some day and surprise you by living far beyond your predictions. If a physician wants to quote statistics, he can give hope to the patient by telling

290

him about unusual survivals, spontaneous regression of cancer, that he will always be there for him to do his best, and that he's alert for new discoveries. The opposite –relating statistics of imminent death to a patient- will cause his body to respond accordingly. We don't want to give false hope to a patient, but when we tell him the truth about his disease we should include supportive statements, and explain to him various forms of possible therapies. Above all, show him love, empathy and support.

Attitude is a quality of the mind, and thoughts can influence the body in either positive or destructive ways.

Doctors and patients should practice projecting in a positive image of survival. It even helps sometimes to keep a sense of humor. Laughter melts barriers between people, and can bring a patient closer to his doctor. Many of my patients, when coming to visit me, would show their cordiality by bringing me a joke that made us laugh. When everything goes well, a good joke brought up during surgery helps in breaking the tension in the operating room. Dr. Cooley is blessed with a sense of humor, always enjoys a good joke, and would carry it from one room to the other.

Laughter is truly the best medicine.

In my experience patients who had vigilance, and were self-assured, had the will to live. People with positive emotions healed faster than others, and could sometimes overcome their disease. It's difficult to heal a patient who has already given up.

CIVILITY AND SERENITY

Serenity is an important quality in inspiring miracle cures. In a serious situation, acting erratically and hastily can result in mistakes. Serene people act with self-confidence and tranquility in times of danger without losing their cool. Being in control in the face of atrocity reflects peace to others. It's a unique and most admirable quality, a trait one can be proud of.

Most of the surgeons I've worked with or observed, even famous ones, sometimes lose their temper during an operation that got tough. It could be out of concern for the patient, but it also reflects a lack of self-confidence. It only serves to create tension in the operating room, and this type of erratic behavior can sometimes lead to mistake after mistake, creating a nightmare for everyone involved.

There is one surgeon who sets a prime example for the present generation. He keeps calm under pressure, uplifts his assistants when they make a mistake, and creates a pleasant, productive and cheerful atmosphere in the operating room. He practically makes surgery fun, and turns difficult operations into simple, achievable procedures. He has positively influenced and helped generations of surgeons, changed operating room procedures for the better, and did it all with modesty. He's my mentor, Dr. Denton Cooley.

When I worked under Denton his daily schedule was 20-25 open heart procedures in five rooms. With two assistants in every room, he performed the crucial parts of the operation himself with an unmatched grace and elegance. One of the great qualities I acquired from him is composure under pressure. It's a unique quality to never crack under tension, and always exude a relaxed confidence. Sometimes he would joke with one of the dozens of spectators who came to watch him in the operating room. They admired the ease and efficiency with which he conducted the surgery and they left inspired. While working in one room, a complication might occur in another, but I always enjoyed watching how he handled such situations with serenity.

Dr. Cooley is also a sportsman par excellence. None of us were up to his standards, and whenever one of us made a mistake, he'd correct it without being abrasive. On the contrary, he'd teach us to learn from our mistakes without humiliating us. This is the essence of nobility.I don't think there has been, or will be, a surgeon with the mental resilience to take on 25 open heart surgeries a day so calmly. Dr. Cooley's efficiency garnered him a high number of referrals from all over the world and an immense reputation among his referring colleagues. He kept them informed about what happened to their patients from the day of surgery to the day of discharge, and in spite of his busy schedule, he answered every correspondence. What's more, he's a lovable man who maintained his charm throughout his success.

It was a great privilege for me to work under Dr. Cooley and absorb his qualities. Much of my success in achieving miraculous cures can be attributed to his influence.

In addition to the surgical training he gave me, I also admired his immense personality.

I am dedicating this book to him, because he sets the image of what a surgeon should be at all times, not that of the angry, pompous, feared professor, but the confident, serene, encouraging, helpful, humble surgeon-in-chief. He managed by example; not by tyranny.

CONCLUSION

Serenity is one of the most valuable gems of character. Serene people do not fuss, get angry or scared. They stay poised, steadfast and calm in the face of difficult situations. They know how to control themselves and interact with others in a tranquil manner. People admire their spiritual strength and feel that they can learn from and rely on them. They're always loved and revered.

HUMILITY

Humility is needed to learn something new. You don't know everything, so open your mind and explore.

Humility is a trait very much needed to be a good doctor, and it's important for every success in life. The paternalistic attitude in medicine that the doctor knows it all, and that knowledge starts and ends at his level, and that he shouldn't be challenged, is, by today's standards, obsolete. We now live in the Information Age, and most patients have a great deal of access to information about their symptoms and choices. With all this information at their fingertips, most patients want to be involved in their own health care. Intolerance to a patient's views is a form of arrogance that stops the doctor from seeking further knowledge. Physicians should not be blindly opposed to new ideas. Very often someone may scorn a new idea that later becomes a success.

A really great man is always humble, and the most humbling experience for a doctor is when he becomes a patient himself. He then experiences the same worries and insecurities as any patient. Any lack of attention from his colleagues becomes apparent and hurtful.

Modesty is a sign of greatness. The Bible says, "Blessed are the meek, for they shall inherit the earth." It also says, "Pride goeth before the fall." Pride and egotism come from a sense of superiority, self-love, or self-righteousness. We're all created equal, but pride can lead someone to unfairly criticize others or find fault in anything they do. This is a subtle form of hatred and violates the law of love.

When a patient asks the doctor about a different mode of treatment or alternative medicine, he should not act as if he has a monopoly on healing, and then chastise him. Rather, we should tell him we'll study that method and offer a balanced opinion. We should never let the public think that we are closed minded to any other form of treatment, especially when medicine is not yet a perfect art, and stops short of curing every ailment.

There's validity in coupling alternative and traditional medicines. It's now known as *complementary medicine*. Doctors should not present themselves as the absolute authority in matters of health and illness. A humble doctor gains new knowledge and accepts alternative techniques, sometimes from his own patient's inquiries. There is a lot to learn from research and the experience of others in curing what is thought at times to be incurable.

We should have the humility to accept others' opinions and suggestions without prejudice, and give them a fair and open minded evaluation. A conceited doctor will repel patients and miss the opportunity to learn from others.

ALTERNATIVE MEDICINES

Under pressure from the public and their demands, the medical profession is having another look at alternative medicine. It's using what's valid and proven useful, and discarding what has no scientific value. The adoption of alternative medicine as complimentary is now common among many doctors.

In the patients' stories, you'll find examples of the beneficial use of alternative medicine, as in the case of Richard H. He refused surgery for his leg claudication and requested to be treated medically using chelation therapy. He knew about its benefits in vascular disease and general health from friends, plus the testimonials of thousands of patients and doctors. Many patients resort to alternative medicine because of what they regard as mechanized medicine, resulting from doctors' over dependence on modern diagnostic techniques, rather than listening to their history and performing physical examinations.

What increases this mistrust is the explosion of medical malpractice whereby trial lawyers attempt to demonize doctors and accuse them of deviation from what they call "usual and customary" or "standard of care." In reality, there is no "standard of care" that fits every patient. These lawyers deprive us of our ability to expand and improvise, and look for better techniques. The worst thing is they can use closed-minded doctors, or even bribe some to act as false "expert" witnesses or "hired guns" against good progressive doctors. Doctors need to

297

believe and pass on the message that medicine is a progressive art that has its limitations, and that it's trying to seek perfection, and isn't totally infallible. We must be honest and admit that every medical and surgical procedure has a percentage of morbidities and mortalities.

Some alternatives are ancient and have stood the test of time, like Chinese medicine, herbal medicine, naprapathy, homoeopathy, chiropractic medicine, chelation therapy, and many more.

Despite this, some doctors look down on such alternative forms of medicine and describe them as "quackery." And when patients attribute some benefits from them such doctors simply describe it as a placebo effect. But even the value of the placebo effect is being more and more appreciated by the academic community.

We now know that when a patient is constantly taking a medication and that it's interrupted to avoid addiction, an injection containing a little bit or none of that medication will prompt the brain to excrete chemicals that give the patient the same effect as the actual medicine. This has been proven by PET scans, other diagnostic modalities, and extensive research.

I cannot forget the experience when my wife had a cluster headache which did not respond to even the strongest analgesics. Luckily, I had a book on Chinese medicine which described a method to stop cluster headaches by massaging the shoulders and neck, and then applying pressure on the temples. Upon using it, my wife's headache promptly subsided. Hence I include in my practice any method that can benefit my patient.

In spite of all the resistance and prejudices, alternative medicine has come a long way. There is now an office of alternative medicine at the National Institute of Health, with reimbursement of selected alternative procedures by some insurance providers. There is a revitalization of books and conferences concerning alternative research and practice, and the AMA has suggested and encouraged hospitals to open departments of alternate medicine. Americans spend billions of dollars a year on alternative treatment options.

While emphasizing the progress in scientific medicine over the last century, we must recognize that our patients possess emotional, spiritual and rational dimensions that are essential in the diagnosis and treatment of disease, and include these factors in our practice. In my own experience, when I have added some of these alternatives to traditional medicine, they've brought about some miraculous cures not obtained by traditional medicine alone. This included prayer and guided imagery.

FAITH AND MIND-BODY MEDICINE

Dr. Michael DeBakey said, "No medical discovery, innovation, or advance is a product of a single mind although the public tends to associate such discoveries and innovations with the name of the one man who popularized them."

The clinical application, however, usually represents a culmination of research and studies by a great many workers, each of whom contributed fragments of knowledge to the original concept.

Mind-body medicine started in the 1850s when Pincheas Quimby, a watchmaker, who watched a hypnotist in a show suggest to the hypnotized person that his arm was burned. Very soon his arm became red and blistered. Once the man was told that his arm was healed, the redness and blisters disappeared. That keen observer asked himself, "If another person's mind can influence our bodies this way, what about our own minds?"

Quimby studied hypnotism and the effects of autosuggestion. He treated patients by explaining to them that their symptoms originated from their beliefs.

Since then, many people in the clergy have published books dealing with how our minds can heal us, how faith can heal, and how we can heal ourselves. Others have treated sickness by affirmations and denials, affirming that we were created perfect and deny disease as an

300

intruder which has no power. Each of these methods emphasizes health repeatedly until our minds accept it as reality and deny sickness as an illusion. And when our minds get convinced of it, our bodies respond accordingly. It wasn't until Dr.'s Matthew and Groopman started writing about the subject that it found its place in medical literature. They popularized the use of hope and faith in curing patients.

Faith is a belief that a patient can overcome his disability, in spite of his disease. Once this message is sent to the brain, it excretes happy chemicals and chemical transmitters which make the body act accordingly.

Primarily, we should practice medicine with compassion, doctoring with the biology of the disease as secondary to the patient himself. We should focus our energy on the total person and not just organs and tissues. We should practice whole-person medicine, treating all aspects of the patient. Dr. Larry Dossey discovered the scientific value of praying and creating the unity of human and divine consciousness that transcends time and space.

It's strange that many pioneers think of the same idea at the same time for different motives. Dr. Matthews at Duke University viewed medicine as composed of a compassionate doctor-patient relationship, which led him to search for a spiritual dimension and practice whole person medicine. He was encouraged to follow this thinking by his mentor, Dr. Engel. The author of Timeless Healing, Dr. Benson, had a different motive, encouraged by his own experience in medical school. Medical schools throughout the world give their students the same basics of medical science, with many schools expecting them to use that information exactly as is. What distinguishes schools like Harvard and Johns Hopkins is that they teach their students to think beyond the basics.

With regards to health, the mind is the master of the body; happy thoughts can bring about youthful vigor, while unhappy thoughts attract disease and decay. Fear can kill a patient quicker than any disease. In fact, fear of disease can often attract disease. We're victims of what we fear, and often the very thing we fear may come to pass. Similarly, anxiety can throw the whole body into dysfunction. Cheerful thoughts of hope and optimism are the best physician's assistance and sometimes are, by themselves, curative.

THE COURAGE TO OVERCOME THE FEAR OF MALPRACTICE

"Whenever there's a chance to save a high risk patient, it's worth taking that risk. Faith destroys evil."

In the story of the Mexican Cowboy, you see that the hospital director and the anesthesiologist were trying to block a lifesaving surgery because the patient was high risk, and the surgery could precipitate a law suit for the hospital. With much effort, I was able to convince the hospital administrators that doing nothing might result in a higher risk for litigation. They allowed the surgery with great trepidation and the surgery was successful, saving the patient's life.

Malpractice in the USA has a negative effect on medicine. Its epidemic proportions are caused by the greed of some trial lawyers and patients who attempt to extort the system, intimidate doctors and hospitals, and inhibit them from using their full capabilities.

At one point, childbirth was done at home by midwives. It was then placed in the hands of general practitioners, who turned it over to obstetricians—everyone dumping it on the other for fear of being sued if a baby is born with a disability. Worse than that, some obstetricians have now abandoned deliveries, because they require unaffordable insurance premiums, not justified by the reimbursement. Thus, they confine themselves to simpler gynecologic procedures.

The negative effect of malpractice does not end there, but they make the sued doctor preoccupied and emotionally upset. This can influence his perception and judgment, and may create errors in his practice, inhibiting him from accepting risky patients. The fact is that without risking failure there is no chance of success.

Surgeons refuse to be involved in complicated or unusual cases and pharmaceutical companies refuse to deliver new vaccines in case one in a thousand might create allergies or complications and give way to litigation. In other cases, some companies withdrew useful medicines from the market—like Vioxx, a wonderful arthritis medication—for fear of class action suits. If one or two patients report complications hungry lawyers initiate a class action suit whereby they collect false testimonies against the product. Some patients fall for it with the illusion they'll get lucrative compensation.

Silicone implants were once labeled as having severe allergic complications, and it became more economical for the company that produced them to withdraw them from the market and declare bankruptcy. But further research proved that the implants were safe, and the promised rewards by the lawyers didn't materialize. The main beneficiaries from those settlements were the hungry lawyers themselves. The bankruptcy of the manufacturer caused hundreds of workers to become unemployed, bringing suffering to their families.

As the proverb says, "A lawyer is a person experienced in pretending to save your estate, but at the end he keeps it for himself."

Malpractice targets the medical profession in an exaggerated manner. It forces doctors and hospitals to pay higher insurance premiums which reflect on the patient and make it impractical for some doctors to practice their specialties because the premiums exceed their income.

The epidemic of malpractice suits in the US is the main reason for the massive increase in health care costs and subsequent decrease in its quality. Almost every doctor practices **defensive** medicine, which prompts him to order x-rays, CT scans, and MRI's and other expensive tests—unnecessarily—in order to cover themselves in case a malpractice suit accuses them of negligence if he didn't do them. It's amazing how medical reform ignored this problem. Some politicians and lawmakers receive lucrative contributions from the trial lawyers lobby to oppose any solution to this problem and minimize its effect on the cost and quality of healthcare.

Beyond that, malpractice inhibits doctors from practicing **intuitive** medicine and listening to hunches, or even trying new ideas which might save a patient's life. It smothers their creativity and confines them to an evidence supported diagnosis. Even then, they can plant doubt in the evidence. They can convince a misinformed jury of anything! Malpractice trials are becoming theatrical. Dramatic trial lawyers are coached to be actors, and they hire false witnesses. Ironically, judges and other supporters approve of this melodrama.

In my opinion, no judge should preside over a malpractice case without taking a *legal medicine* course. And if the jury is truly composed of your peers, then it must be composed of neutral doctors.

The so-called "informed consent" required a doctor to communicate to the patient every possible complication, no matter how rare. According to mind-body medicine, such negative predictions stick in the patient's mind, delaying healing and prolong recovery. Some patients might be scared into refusing surgery because of such negative predictions, and die from their disease without benefit of intervention. Because this method ignores the major benefits from success, and concentrates mostly on the small possibility of complication, I can't help but wonder if this shouldn't be considered misinformation. Doctors can't work efficiently when they feel a dagger is pointed at their back. Defense attorneys, on the other hand, are more objective, honest, and medically informed. They're usually paid less fees by insurance companies, but some of them feel bad about the stress, humiliation and loss of time doctors are exposed to because of litigation. Generally speaking, they are less aggressive and much more academic and truthful. The expert witnesses hired by defense counselors are more honest, but lack the theatrical vehemence of the highly paid plaintiff's witnesses, who often lie to win. Such so-called expert witnesses often make it a career, taking courses in it!

Plaintiffs' expert witnesses are even given immunity to lie. They charge up to $16,000.00 to appear in court. Such exorbitant sums lure them into giving false expert opinions. They prostitute themselves, besmirching their profession.

ADVICE FOR AVOIDING A MALPRACTICE SUIT

In the litigious atmosphere trial lawyers have created, I cannot imagine that a health provider would go through his career without at least one adverse event. Only those who don't do much end up avoiding lawsuits. If a complication arises, the best thing to do is immediately report it to the patient and his family. Explain why it happened, and how it will be corrected. Legally, if a complication occurs during an operation, and you correct it, you cannot be held liable. You can apologize, but without admission of guilt. Never blame another physician for what happened. It may prove to be inaccurate as things move on. If a complication happens in a hospital, have a meeting with the patient and his family members and the healthcare team. Report it to the hospital's risk manager and to your insurance carrier. Always have all the facts in writing on your personal and hospital records.

OTHER STEPS TO AVOID BEING SUED

1. Obtain consultations, because consultants are considered experts in their fields, and it's hard for trial lawyers to find experts to contradict all of them. This decreases their chance of winning.

2. Report to your risk manager and malpractice insurer; they will give you advice on how to handle the situation, and start the process of gathering information to protect you. Lawyers are reluctant to attack a supported doctor. This will give the patients an understanding of the circumstances of how an error could have happened and how you have corrected it. It leaves no harm, and even if it does, it leaves so little harm that it doesn't merit a law suit.

3. Make arrangements to hold a follow up meeting with the patient, his family, and the hospital team.

4. Follow up with other meetings, and assure the patient and his family that you'll always be available to help. Also, answer any inquiries they may have.

A good doctor-patient relationship can avoid a claim if handled with compassion and communication. Spend as much time as possible to be courteous when dealing with the patient and his family. It may be appreciated if you or your hospital waive or decrease charges out of courtesy "without admission of guilt." That might also save you.

WHAT NOT TO DO IF YOU ARE SUED.

1. Stop communicating with the patient or his attorney.

2. Never transfer property under your name to someone else.

3. It's very important to have your insurance company appoint a lawyer you can trust and who has a good reputation. A lawyer who has a pessimistic view of your case can do more harm than good. He's employed to defend you, not accuse you.

4. When you receive a formal complaint don't let exaggerated allegations shake you. They're more formalistic than real, and your attorney can negate them all.

5. When you meet with your lawyer, make sure he's acting on your behalf, and not merely in the interest of the insurance company. And

have a complete copy of your reviewed
medical records available for him.

6. Its upsetting and emotionally disturbing
   when the complaint and summons arrive at
   your door with the greatly exaggerated
   accusations of wrong doing. As Mr. McCarthy
   describes in his book, *Malpractice Cures*, these
   are generally worded as one size fits all, so
   don't take it to heart. Contacting your
   insurance carrier will ease some of the burden.

7. Do not feel humiliated because of the
   allegations raised against you. Be optimistic
   that the case will be resolved in your favor.
   If the situation takes a stressful toll on you,
   your insurance can arrange a psychological
   counselor to support you.

8. Before you give a deposition, make sure your
   lawyer has deposed the plaintiff's witness
   first. The defense you make against these
   allegations may be used in the process called
   discovery. Be sure that you covered, in your
   deposition the defense you will use in court.
   Anything you do not say in the discovery you
   cannot use later in your defense. Plaintiff
   lawyers might take advantage of this and not
   let you use a defense discovered during the
   trial which you did not share during the

deposition. Also, mention all the references you might use. If you don't, you cannot refer to them in your defense.

9. Your defense firm might have competent experts to be on your side. If not, refer them to prominent experts in your field.

10. If the case is filed after the Statute of Limitations, your lawyer can file a *Motion to Dismiss*. The plaintiff can do the same if he finds that his client is lying or has no valid grounds to go forward with the case.

11. If the balance of the depositions were in your favor, the plaintiff's attorney might ask to settle. With your approval, your insurance company might agree to a settlement to avoid the cost and strain of a trial. A few states like Indiana offer arbitration or mediation. Though settlements take a strain off your back, they're now being reported to a "data bank" and you can be questioned by your state's board of medicine for a period of up to ten years. They're by no means a measure of any doctor's competence, but could be used against him.

12. Don't let your lawyer ignore any of the plaintiff's witnesses, even if they seem ridiculous. For example, a financial representative usually

relays to the surviving members of a family imaginary tables and predictions of the degree of loss that happens as a result of the natural death of the patient.

A final figure if not challenged, will remain in the jury's mind. If the patient has a family history of sudden death at a young age, your lawyer should not accept the calculation as, "If he lived to be 70, his income would have been so and so." Your lawyer should object to this every time the plaintiff's lawyer describes an inevitable death as a *wrongful death*.

PRAYER AND MEDITATION

"The response to prayer comes out of the creative mood that it creates and not by the words. Words by themselves do not heal, but it's the feeling that flows through the words that heals."
—Charles Baker

In the first chapter of this book, we discussed Jeremy, a prayerful man with a malicious cancer who was cured by a miracle, probably brought on by his faith and prayer. People affected with his malicious cancer typically die within two years, no matter what treatment they receive. What led me to perform a potentially dangerous operation on him were his faith and resilience, which I am convinced, helped him live a full and productive life afterward. God truly works in mysterious ways when someone believes.

I define prayer as the way we communicate with our Creator, and meditation as the way we listen to Him. By listening to God, we're guided and blessed.

At the turn of the century, the mystic Joel Goldsmith promoted a type of prayer he called practicing the presence. In this, we become aware of an omnipresence that follows us wherever we go and in whatever we do. It protects and watches over us during an operation or while we're treating our patients. He also explained what is meant by prayer without ceasing. In this positive, loving state of mind we feel a oneness with God all the time. It guides us in all our activities. We'll be blessed, and our needs will be met, even without asking.

The Bible says, "Before you call, I will answer, and while you are still knocking, I will hear." "What things you so ever desire, when you pray and believe you shall have them."

What God expects from us is to "seek His kingdom and its righteousness, and all other things will be added unto us." Personally, I always pray to be worthy of His kingdom and try to be aware of His presence, His love, His power, His oneness and His protection. And I'm always grateful to Him for answering my needs as well as those of my patients.

I never pray for material things; they come by themselves. When I reach that state of awareness and bring my patient into the picture, his recovery is very much enhanced. (See my Complete Prayer of Protection with James Dillet Freeman at the end of this chapter.)

In the prayer of affirmation and denial, we affirm that God is love and His love will give us the ability, knowledge, wisdom, and courage to heal. We deny anything that contradicts these beliefs.

Faith is the belief of things happening before they appear or are perceived by human senses. That's why we say, "Thank you God for answered prayers." Prayers of gratitude are very rewarding.

I also use visualization. I close my eyes and picture myself and my patient in a healthy state. Eventually it happens, and according to the faith of my patients, I put them in the presence of their Lord. Sometimes I visualize them surrounded by light and its healing energy flooding their body. Putting myself in the light and sending it to others brings a great blessing to me and healing energy to them.

When I meditate, I relax and free my mind of all worldly affairs. According to the relaxation response this lowers adrenalin and cortisone levels, as well as blood pressure. Sometimes I hear God's voice tell me something or reveal part of His nature to me.

I believe medicine is not the only way of healing but it is the most popular and practical. All healing comes from God in different ways and through different people, and I myself am merely a vehicle through which the knowledge, talent, and ability He gave me can be used to heal the patients He sends me.

I never advertise. Many of the patients I encountered were through circumstances I never dreamt of—as if God was directing me to them or them to me. We'll discuss this more in the chapter about coincidence and destiny.

One of the most effective prayers which I often read with my patients, or give them in a pamphlet, is the 23rd Psalm. "The Lord is my shepherd, I shall not want; He makes me lie down in green pastures. He leads me beside still waters; He restores my soul. He leads me in paths of righteousness for His name's sake. Yea, though I walk through the valley of the shadow of death, I will fear no evil. For thou art with me. Thy rod and thy staff comfort me. Thou preparest a table before me in the presence of my enemies; Thou anointest my head with oil, my cup overflows. Surely goodness and mercy shall follow me all the days of my life, and I will dwell in the house of the Lord forever."

Although prayer is a spiritual way of healing, its benefits have been scientifically proven by many of my colleagues.

I could not close this chapter without mentioning the Quantum Theory discovered by Albert Einstein. He concluded that energy and matter are the same. Since prayer and light are energies, they can produce objects or actual form. Hence, prayer can be transformed into healing that becomes reality. This scientifically explains the power of prayer which has been proven by double blind studies.

The Complete Prayer of Protection

The light of God surrounds us.
The love of God enfolds us.
The power of God protects us.
And the presence of God watches over us.
Wherever we are, God is, and all is well!
The Mind of God guides us.
The Life of God flows through us.
The Laws of God direct us.
The Kingdom of God abides in us.
The Joy of God uplifts us.
The Strength of God renews us.
The Beauty of God inspires us.
The Glory of God is reflected on us.
The Abundance of God prospers us.
The Grace of God is our sufficiency.
The Peace of God relaxes and comforts us.
The Passion of God stimulates us.
The Perfection of God perfects us.
The Wit of God answers for us.
The Vision of God expands our imagination.
The Humor of God laughs through us.
The Energy of God energizes us.
The Timelessness of God rewards our patience.
The Faith in God reassures us.
The Infinite memory of God is stored in us.
The Oneness of God is the truth that sets us free.
We are God's Expression in words, acts and deeds.
Nothing is lost in God's universe.
Wherever we are, God is,
And all is well!
Thank you God for your love, protection,
And creating us in Your own image.

-James Dillet Freeman & Joshua David Salvador

The first four lines of the *Complete Prayer of Protection* were written by Mr. James Dillet Freeman and were given to the astronauts who implanted them on the surface of the moon. It will stay there forever. (No eroding atmospheric wind or rain exists on the moon.) The rest was inspired to me in moments of contemplation, and given to the Unity Churches. These verses hang on their walls so the congregation can read them and get inspired. Sometimes, when a patient needs a prayer, I read it with him and give him a copy. They direct the attention of a scared patient to the protection of a higher power, and to many they've given comfort.

MIRACLES OF SPONTANEOUS CURES

"He who doesn't believe in miracles is not practical."
—David Ben Gurion

In order for miracles to occur, you must believe in and have a positive expectation for them.

Miracles come from love, are created by love, and they are attracted to us through love, especially when you believe in and expect them. Until recently miracles were regarded as gifts from God and impossible to achieve by ordinary people. The thinking is, "what's impossible for man is possible for God," meaning they're good acts that can come only from God's mysterious powers, but cannot be explained by science. The medical examples of such miracles are spontaneous regression of disease and tumor necrosis, and such occurrences were documented by medical doctors. From a religious point of view, miracles are instances in which a supernatural power modifies the natural world. From this perspective, the limited human brain cannot explain the actions of the limitless power of God. His wit, reason and intelligence are beyond human comprehension. Yet, the last fifty years have revealed miracles produced by man as a result of his thinking, imagination and creativity with which God has endowed us. Many inventions reflect this, including the discovery of flight, followed by space flight, then by man landing on the moon, and recently a camera on Mars and a city suspended in space. Isn't that similar to the Earth which is suspended in space as a planet among others, also with no pillars? This all includes man's creative power, because man was created in the image of God. It's possible then to say, "It is all God."

In medicine, miracles include heart and other organ transplantations. Who could imagine that one day a person could live and walk with another person's heart; or the ability to create a human being in the image and likeness of another human being by cloning one of his cells, has met disapproval as a moral and religious issue, yet the ability to create it is no less a miracle.

The next medical miracle is the discovery of the stem cells which, if injected in any organ, can regenerate it. The discovery of the genome has resulted in removing the genes of some diseases and is promising a cure for many others.

The discovery of the DNA has saved the lives of many people condemned to die by capital punishment, because it proved them not to be the guilty criminals.

From all this we conclude that miracles, which were once attributed to God, through the recent partnership of science and spirituality, open the doors to science to create human miracles. The science of quantum physics, which proved that energy and material are one, means that any thought up energy can manifest as a discovery. That means that, what was considered supernatural yesterday is natural today. Today's miracles are the result of a series of natural events occurring in the right sequence and at the right time to produce wonderful results.

According to the Law of Mind, or the law of co-creation, consciousness manifests. Consciousness is the summation of all our thoughts, beliefs, feelings, desires and intentions. The more we're able to align our thinking with what we want, the more we become fully empowered to create miraculous discoveries. This is so, because:

1. Every one of us is a thinker in our world.

2. We choose the thoughts that help us create something of our thought patterns.

3. Our thought patterns manifest in our lives.

4. We should always create thought patterns that establish our good.

5. Our good is whatever we desire in life.

6. Once we accept the idea that we are entitled to have good things happen in our lives, the door to manifest that good becomes open.

7. The universe supports us in bringing good manifestations into our lives.

This needs for us to believe that there is a Power and Intelligence making big miracles happen. The attitude is that Life, Intelligence and Love could make good miracles happen. We have to keep an open mind, because our mind is the instrument of our miracles. No miracle is too big to manifest. You have to believe that you're already wired for miracles. You don't need to focus on how or when. Focus only on what you want to create, and not what somebody else wants. And make your desire so clear in your mind that you can expect it to manifest at any time. We need faith in ourselves, and an open mind to all possibilities. Keep a good relationship between yourself and your spirit. Spirits will work for you by working through you. Pray often because prayer works.

THE SOURCE OF MIRACLES

So, are miracles the products of God's spiritual powers or scientific human achievements? Let's ask a scientist.

The greatest scientist of our time is Albert Einstein. He states, "Science without religion is lame; religion without science is blind." This thought merges science and religion, which evolved into quantum phenomena of miraculous character. They are forms of a synchronous experience.

It was Einstein who proved that energy and matter are one; they are interchangeable. Every material thing we see on this Earth is a product of an idea. Ideas are in the realm of the invisible; they're a product of the electro-magnetic forces of the brain. They're referred to in the Bible as *The Word*. "In the beginning was the word and the word was God." When ideas materializes, the formless takes form. Since ideas are a form of energy, everything in the world comes from the void called energy, e.g. atomic energy.

The atom has no shape. It cannot and will never be seen by any magnifying power. It's formed by a positive nucleus, and negative orbits around it. Our body is formed by cells and cells are formed from millions of atoms. Every cell has its personal intelligence stored in its RNA which, when transformed into DNA, results in an explosion of energy. Do you now see how ideas or thoughts can influence human molecules and organs?

This brings us to the miracle of mind-body medicine. Our thoughts influence our health. Positive thoughts bring healing results, negative thoughts bring destruction. If we excel in the former, it becomes a remedy which we can carry with us all our lives and it can be used to bring good health in all aspects. Most diseases, no matter their cause, come to us in periods of distress and emotional depression, either by lowering our resistance to infection or creating poisonous products like cortisone, insulin, and adrenaline in response to the stressful situation.

Ayurvedic medicine creates a positive attitude in both patient and doctor, and has been proven to bring about positive healing results. This type of medicine has been popularized in this country by Dr. Deepak Chopra. I highly recommend his book *Quantum Healing*.

The spread of mind-body healing is one of the greatest discoveries that can result in remarkable cures—miraculous cures.

In their book, *Shortcut to a Miracle*, Michael and Elizabeth Rann give us the keys to making miracles. Love, forgiveness, faith and optimism are essential factors in their production.

To receive a miracle cure both doctor and patient should anticipate it happening. If a patient has an incurable disease, and his doctor is loving, compassionate and determined to heal him, he will be inspired toward a way to accomplish it.

We're all connected. All we have to do is become receptive and tune in to God's voice and universal mind.

COINCIDENCE OR DESTINY

Upon reviewing some of those miraculous recoveries, I recall that some of them were emergencies that neither were on my schedule nor were they my patients. Was it that I happened to be there by accident in the right place at the right time? When this happened more than once, I started searching for an answer.

Could it be that somehow these patients were attracted to me or me to them by some mysterious life force, a deep call for help and I merely was the closest and sent to respond? Or was it just by accident?

Five cases stand out:

- The young girl who was pinned under the car.

- The landlord stabbed by his tenant over a rent collection dispute.

- The patient who walked in with an amputated arm.

- The nurse who was shot in the face with BB gun pellets while leaving the hospital.

- The woman with massive pulmonary embolism a month in coma.

Let's analyze the circumstances that brought these patients under my care:

The first case happened just as I was leaving the hospital in the afternoon. I could've left sometime before or sometime after, and don't know why I left that particular moment. Was it a call to help or a mere coincidence?

I stumbled upon the stabbing victim while leaving the hospital through the Emergency Room, rather than by the usual exit. I noticed a man with a chest tube draining blood with a normal saline IV instead of a replacement transfusion. It seemed like an unusually serious situation, which made me inquire who was taking care of him and why he was not receiving a blood transfusion. The patient's brother, a lawyer, was within earshot and demanded I give his brother surgical care. There was little time between life and death when I took him to surgery and ordered the blood bank to transfuse him with O blood which any patient can receive, ignoring the possibility of incompatibility and transfusion reaction.

In the third case, the patient walked into the emergency room with his forearm severed, attached by a piece of skin and a tourniquet. A couple of weeks before, I had attended a meeting of our Denton Cooley Surgical Society during which Doctors Zaorski and Bangash presented a similar case and described the technique of reattaching a severed arm. Was God preparing me in advance for this case? After all, God is omniscient and knows His people's destiny before it happens. The circumstances beg the question: coincidence or destiny?

In the fourth case, the woman shot in the face with BB pellets, it landed on my lap. Her type of trauma had not been described in medical literature, and nobody else wanted to take the responsibility, so it became mine.

In the fifth case, where the lady was in a coma and had been given just three days to live, Camacho asked if I could help his partner whom I'd never met before. I had her transferred from Vencor to my hospital with immense opposition from the President and CEO. The other hospital noted they had done all they could to dissolve the blood clots in her lungs. When I stumbled on her case, they had given up hope. On the tenth day, she was out of the hospital and a few months later my nurses invited her to my sixty-fifth birthday party where she danced for three hours.

The coincidence occurred when her husband was driving with one of my patients and they called me five minutes before I left the office. I don't know what made me accept the risk and taking the responsibility of a dying patient I had never seen before.

COMMENTARY

While we are alive our brain emits electric waves which are recorded by the EEG (electroencephalogram). These brain waves vary in intensity and format, according to the demeanor of the person. At the same time, our brains are capable of receiving brain waves directly from other people through ESP. In the case of an emergency, the person in need projects calls for help waves in an accelerated form and in proportion to the degree of danger. The closest person capable of handling the situation receives these brain signals first and responds to them. We refer to these respondents as angels.

We live in an interconnected universe and I believe nothing happens by accident. We are directed, and we are being called. Similar situations can happen to anyone of us in our lifetime. I recall being in similar situation myself at least twice.

The first instance happened when I was driving on a highway at a high speed from Tel Aviv to Haifa where my wife, some friends and relatives were waiting for me for a Sabbath dinner. Suddenly, I lost control of the car. I didn't know what happened, but the car spun out twice on the highway. Right at that moment I said "God, I cannot control the car. I don't know what happened? Please take over and help me to control it." Luckily no cars were behind me at that moment. Finally, the car stopped in a sandy area next to the highway. It was dark. I turned around and was in a state of shock and disbelief. A feeling of helplessness and fear engulfed me as I looked around at the empty space surrounding the highway. I didn't have my cell phone with me, and I didn't know how I'd arrive at my destination nor how to contact my family or even call for help.

Within a few minutes, out of the blue, a car came straight for me. It stopped, and a father and son got out. The father told me I had a flat tire and asked if I had a spare. Before I could even answer, the son lifted the car with their jack. He took off the flat tire and the father replaced it with the spare. Then he told me, "Get into your car and start the engine." The car started easily and then the father added, "Drive the car around. You're safe now."

They waited until I drove around and before I could even say thank you, they went back to their car and drove off.

I don't know where they came from. They just appeared like angels and saved me from a vexing, overwhelming situation.

The second instance happened to me in Chicago. I had an important meeting downtown, and as I came to the street where the meeting was going to take place, I found that it was a one way street in the opposite direction. I told the traffic police woman that I was late for an important appointment, two buildings away from the intersection and asked her permission to let me drive that short distance since the street was empty from traffic. I didn't even know if she heard me, but I think she thought I was disobeying her authority. In a split second I saw police cars coming toward me from everywhere. Apparently, she radioed them regarding my violation.

About a dozen policemen got out of their squad cars and angrily pounded on my window, demanding I come out.

"You're under arrest" one of them said. "Open the door," shouted another. I was apprehensive as to what they were up to because I could see extreme anger in their faces. I locked myself in the car and they tried to force my door open. I attempted to call someone from my car telephone but I couldn't get a connection and didn't know whom to call. "There goes my meeting," I thought. "They'll think that I didn't show up intentionally."

I asked God to help me in what was a serious, scary situation. Suddenly, a gentleman in plain clothes came out of his car and showed me his identification through the glass window of my car and said, "I'm the police captain. Don't be afraid. You can open the window."

When I looked around and saw that the angry policemen suddenly calmed down and stood motionless, I trusted him. I opened the window and introduced myself to him. He told me that it was his day off, but as he was driving by, something told him to turn on his radio to see if there was anything important happening in the city. I explained to him that my appointment was two buildings away and I thought the policewoman had given me permission. He asked one of them to stay and issue me a ticket, but then let me drive to my destination peacefully. He told the others to disperse, and they got in their cars and promptly drove away.

To me it seemed as if that captain descended from heaven and saved me from a very scary situation. Isn't it strange that it was his day off, and in that particular moment he turned on his radio out of pure curiosity and heard what happened? Was that by accident, in response to my call for help, or divine intervention?

Whenever I'm in a serious situation, I always say, "God, help me" and expect help will come. Indeed, there's hope for all of us in serious situations. Sometimes I was meant to give help to someone, and at other times to receive help from somebody. If it happens to you, don't panic. Act with faith, knowing that help is near.

SPIRITUAL HEALING: A REALITY WORTH EXPLORING

Spiritual healing dates back thousands of years, whereby prophets and other holy people cured through mystical powers. It was revived in 1850 by people who were cured by it, and then tried to teach it to others. Today there are centers of spiritual mind healing all over the world. A lot of literature and books about it are now available. Examples are the *Science of Mind*, *Creative Thought*, *The Daily Word* and *Unity*, which reach millions of people who believe in it. It has been pioneered by Religious Science, Unity, Divine Science, Christian Science, and Science of Mind. All of them emphasize the discipline of changing thinking from the negative to the positive by understanding that everyone has been created perfect, spiritual and divine.

Although spiritual practitioners don't know the medical names of disease, they are not opposed to seeking out the services of M.D.s. They believe that all healing comes from God, using different people in many different ways.

Pure spiritual healing comes from the Creator's unseen omnipotent power. To assume this awesome power, the healer usually lives, moves and has his being within it. Such healers live in this world, but they are not of it. They pray unceasingly. That means they practice the presence every moment of their lives and meditate regularly.

331

Spiritual healers live in the spiritual dimension as opposed to the material universe. To practice spiritual healing, the healer must rise above the level of appearance, above the corporeal and personal sense, to a higher plain of consciousness, and make room for the spirit of God. Upon acquiring this spiritual energy from the Supreme, he then transfers it to the patient.

The spiritual dimension of the world, or the dimension of the unseen, is waiting to be discovered. Our current knowledge of it is perhaps equal to our knowledge of flight or aerodynamics one hundred years back. To be in the spiritual dimension is to rise above what we see, hear, touch, taste or smell and tune into the unseen spirit of our Creator. It emphasizes the power of God's divine love and caring for us, and denies the power of the condition which afflicts us with sickness. It is affirmation that God is all in all, and that matter and sickness are of no power. Besides God, there is nothing else. It emphasizes the oneness of God and man and the awareness of it.

The Lord is the greatest spiritual healer ever known. How then does the healer know if he has attained this consciousness? He knows by feeling an inner impulse, a whisper, or an inner feeling of being a transparency of Godly activity. In the presence of the Lord problems are illusory. There is no place for fear in the presence of His love. The healer must convince himself that God's power is unique, and errors have no place in that realm.

The climax of the prayer comes when the practitioner feels he is a purely spiritual being. God is within him, and he is an instrument for God and God's love. His body and that of the health seeker are the very temples of the living God, and God is the soul of their being, so that the patient and Father are also one. The practitioner loses the material sense of life. When our consciousness is lifted to that sense, and we feel communion, we become grateful for the prayers he has answered. And in that feeling there is a tranquility of the soul. It's a presumption that what you prayed for has already happened, but will manifest itself in God's time.

These feelings are transmitted from God to the healer to the patient without reservation. A practitioner not only knows, but feels that with God all things are possible.

Some people, like Dean Kraft, are gifted with such powers. Kraft was scientifically tested at Stanford University and was able to heal one of the staff members of cancer simply through his miraculous power of touch. The mystic Joel Goldsmith was able to heal people who communicated with him by convincing himself of the nothingness of illness. It is reported that those patients who wrote to him, "...the hem of his robe..." and would be cured of their illness. Could these miracles be attributed to the power of conviction on behalf of the patient?

NOTE

The results of medicine are not one hundred percent, but at the present time, they are the most effective. I'd like to advise the reader to start with medicine and make use of its great advances. If your treatment is not successful, then seek other alternative methods, such as chiropractors, Chinese medicine, acupuncture, etc. Spiritual healing is not one hundred percent effective, either, but the patient should keep praying and believing in God's power to heal. This is especially true for newly discovered cancer. The cancer should first be removed by surgery, and then seek other methods. God assigned healing and gave the wisdom, ability and talents to medical and surgical practitioners. After that, you can thank God and pray for complete recovery. The important thing is to never lose hope.

IN CONCLUSION

Spiritual healers believe there are two opposing views of the universe:

The **Material Universe** is based on how we perceive the world through our senses, mind, and scientific discoveries.

The **Spiritual Universe** is based on the unseen universe, which we cannot feel, touch or hear, but nonetheless exists around us. Since God is spiritual, and the spiritual is unseen, this means God's Universe has not yet been revealed, but is waiting to be discovered. Because we are created in God's own image, we are spiritual beings. We are all connected by the spirit of God and not by religions or races. Because God is love, we are not only loving people, but we are love *itself.*

The spiritual universe is extremely vast and we've not even scratched the surface in our discovery of it. At present there are many testimonies by people who were cured by spiritual people who live in the spiritual universe and describe themselves as "living in this world, but not of it."

People who acquire such illumination, develop it by a long study of the Bible and continuous meditation. They claim that awareness or consciousness of the universe comes to them from communicating with God. They obtain their power to heal by aligning themselves with His awesome powers and describe Him as omnipresent, omnipotent, and omniscient.

ABOUT THE AUTHOR

Dr. Joshua Salvador wrote this book after fifty years in the practice of medicine, forty of which were spent as an innovative and intuitive surgeon in Chicago. He was so loved by his patients that even after his retirement, he is constantly called by his patients, asking about him, and consulting with him.

His ambition to increase his knowledge about his profession took him to five countries in five continents, where he trained and practiced in most of them.

Through his presentations and introductions of new methods in different parts of the world, including Egypt, Sudan, England, Canada, and the United Stated, gave him an international reputation.

One of Dr. Salvador's greatest achievements is serving at the Military Hospital in Tel Aviv during the Six Days War in June of 1967 where he salvaged injured Egyptian soldiers from Sinai and gave them the same care as any Israeli.

Another great honor after the Six Days War ended was to take care of the great hero General Moshe Dayan whereby he was injured in historical excavations, which was his hobby.

Dr. Salavador is board-certified in general and toracic surgery in Israel and Canada and was honored by being elected a fellow of the Royal College of Surgeons of Canada.

In 1968, he was granted a Fellowship in Baylor University, Texas where he was a Fellow under Drs. Michael DeBakey, and Denton Cooley at Methodist Hospital. When Dr. Cooley founded the Texas Heart Institute next to St. Luke's Hospital, Dr. Salvador followed him and got trained by him, learned a lot from him, and participated in forming the "**Denton Cooley Cardiac Surgical Society**," composed of the hundreds of surgeons from all over the world who came to train under him.

It was a great honor when Dr. Salvador was elected to be the **tenth president of the society**, which at that year had its international meeting in Athens, Greece.

Following that, he spent a year with Dr. Albert Starr in Portland, Oregon. Dr. Starr was the first surgeon to insert an artificial valve in a patient.

He moved to Chicago in 1971 and made it his home, practicing and teaching surgery there for forty years, and became president of the Heart Lung and Vascular Institute of Chicago.

Dr. Salvador is a member of many professional medical organizations and societies, including the AMA, the Society of Thoracic Surgeons, the Society of Chest Surgery and the American College of Cardiology. He also studied the new science of anti-aging medicine, got certified in that field, and became a life member of the American Academy of Anti-Aging Medicine.

When Dr. Salvador retired in 2006, he did not lose his ambition for knowledge. So he studied legal medicine, and medical-legal counseling, and was certified as an independent medical examiner. He is an active teacher and lecturer and has been invited by various organizations to speak.

At the present time, Dr. Salvador lives in Florida with his wife Patricia, R.N. His son David lives nearby with his wife Kathy and their children, Bella, Sophie, Mia and Emma. Joshua also has a son, Arik, in Hawaii; and a granddaughter, Lola, in Holland; a daughter, Katy, in Chicago; with her son, Michael. He loves them all very, very much.

Dr. Joshua Salvador

PERSONAL MESSAGE

Whenever you are in a desperate situation and your doctor says, "we did all that we can, there is nothing more that can be done," look for another doctor who believes, *"there's always another way."*

Have faith: many new medicines and techniques are always being discovered, and one of them may solve your problem. You can get help from alternative medicine, prayer, and belief. With all these methods plus medicine, it is my prediction that there will come a time when *nothing is incurable*.

I am sending you the healing power of light and love, energies which will help you and pave the way for your recovery and I hope the the knowledge you gained from this book will bless you and guide you.

To my patients who wanted their miraculous cures published to give you hope, we are all grateful.

To my medical colleagues, I urge them, beyond giving their patients the best in their knowledge when the situation is desperate, to try to find a new technique to salvage them. That's how medicine progresses towards perfection.

If you have any comments or questions, or a miraculous cure you would like to report, I would be glad to hear about it and include it in a future edition.

You can contact me at my email address:

joshsalmd@aol.com.

POST SCRIPT

A reader once asked me, "Why did you take risks?"
My answers are:

1. I value human life and I would do anything to save someone, regardless of any consequence.

2. Without taking risks, there can be no successes.

3. Every great achievement in the world was done by courageous people, who took risks and set an example for us.

4. I had self-confidence and faith that would help me. I believed that I was destined to be in a place and time where my help was needed, and that He who drafted me for this job wanted me to save a life or cure someone.

5. I am, by nature or upbringing, a selfless person. I cannot see any person in need of help and *not* stretch a hand to him, whether medically, socially or financially, to the best of my ability.

6. Upon entering Dr. Cooley's operating room the first time, I saw a plaque which rang a bell in my mind:

"The credit belongs to the man who is actually in the arena, whose face is marred by sweat and blood, who strives valiantly, who errs and comes short again and again because there is not effort without error and shortcoming, who knows the great enthusiasms, the great devotion, spends himself in a worthy cause, who at best knows in the end the triumph of high achievement, and who at worst, if he fails, at least fails while daring greatly, so that his place shall never be with those cold and timid souls who have never tasted victory or defeat."

—Theodore Roosevelt

GLOSSARY

Abdominal peritoneal resection: resecting the lower part of the colon and rectum for a rectal cancer usually resulting in a permanent colostomy.

Adeno.: Gland; glandular.

Anastomosis: the surgical connection of blood vessels or organs, e.g. bowels.

Aneurysm: an abnormal blood-filled dilatation of a blood vessel and esp. an artery resulting from disease of the vessel wall.

Anoxia: hypoxia; lack of oxygen that could result in permanent damage.

Apnea: transient cessation of respiration.

Appendicitis: inflammation of the appendix.

Arterio: something related to an artery.

Arteriograms: a radiograph of an artery made by injecting an opaque dye in it.

Artificial Grafts: synthetic tubes used for connecting blood vessels.

Atelectasis: collapse of the expanded lung or part of it as a result of blocking air entry to it e.g. by a intrabronchial tumor.

Atresia: closure of a natural passage in the body.

Auscultation: the study and detection of sounds by use of a stethoscope.

AV bundle: a bundle of fibers in the heart for conducting impulses that result in heart beats.

AV fistula: a connection between an artery and a vein usually as a result of injury but rarely congenital.

Axilla: the space between the arm and the chest wall just below the shoulder (armpit).

Axillary tail of the breast: part of the breast closest to the axilla.

Bacteremia: presence of bacteria in the blood.

Basal rales: abnormal sounds accompanying the normal respiratory sounds in the base of the lungs.

Bilharziasis: (schistosomiasis) a severe endemic disease in the Nile and Asia & South America caused by a worn which penetrates the skin during swimming in infested water and settling in the bladder causing severe hematuria.

Bleb: a large air sac on the surface of a lung.

Bleomycin: a mixture of glycoprotein antibiotics derived from a streptomyces and used in the form of the sulfates as an antineoplastic agent.

Bovie: an electric machine that can coagulate a bleeding vessel to stop bleeding during surgery.

Brachial Plexus: (BP) a complex network of nerves that is formed chiefly by the lower four cervical nerves and the first thoracic nerve and supplies the chest, shoulder and arms.

Bronchoplasty: surgical connection between bronchial segments.

Bronchoscopy: visualization of the bronchial tree by passage of a bronchoscope.

Cannula: a small tube for insertion into a body cavity, duct, or vessel.

Carcinoid: a benign or malignant tumor arising esp. from the mucosa of the GI tract or the bronchial tree (lung).

Cardiomegaly: enlargement of the heart.

Cardiomyopathy: disease of the heart muscle.

Carditis: inflammation of the heart muscle.

Carpal Tunnel Syndrome: a condition caused by compression of the median nerve between the Retinaculum and the bone (the carpal tunnel) and is characterized by numbness, pain and weakness in the hand and arm.

Cellulitis: diffuse infection that occurs under the skin.

Cervical Rib: a supernumerary rib sometimes found in the neck above the usual first rib.

Chelation: the use of a chelator as EDTA to bind with a metal in the body to form a chelate so that the metal loses its toxic or physiological effect.

Chordae Tendineae: the delicate tendinous cords that attach the edges of the artrioventricular valves of the heart and to the papillary muscles and serve to prevent the valves from being pushed into the atrium during the ventricular contraction.

Claudication: pain in the calves of the legs on walking.

Coenzyme Q-10: an enzyme that possesses antioxidant properties and facilitates chemical reactions in the cells.

Colonoscopy: endoscopic examination of the colon.

Commissure: a point or line of union or junction between two anatomical parts.

Commissurotomy: opening the adhesions between the two cusps of the mitral valve.

COPD: (Chronic Obstructive Pulmonary Disease) chronic disease of the lung characterized by continuous coughing and sometimes blood in the sputum. This is also referred to as emphysema and very related to smoking that destroys lung tissue and turn it into empty cavities.

Cryo: treating some lesions of the skin by freezing them using extreme cold.

CT scan: computed tomography means turning regular x- rays by the computer into cross-sectional pictures.

Debulking: removing part of a tumor or shrinking it to make it easier to treat by surgery or make it more susceptible to chemotherapy and irradiation.

Diuresis: an increased excretion of urine.

Doppler: a form of ultrasound that picks up blood flow in the body.

Ductal: made up of ducts; bodily tube or vessel esp. when carrying the secretion of a gland.

Dyspnea: difficult or labored respirations.

Echo or **Echocardiogram**: sound wave test of the heart

Embolization: the process by which or state in which a blood vessel is obstructed by the lodgment of a blood clot.

Emphysema/COPD: *see COPD*.

Empyema: the presence of pus in the membranes that surround the lungs in the pleura. (Pleura is the outer lining of the lung.)

Endotracheal: intubation-insertion of a tube inside the trachea to facilitate breathing or for the purpose of anesthesia.

Enucleating: scooping out a tissue or tumor from inside of its outer capsule.

Epstein virus: a virus causing chronic fatigue.

Esophageal cardio myotomy: (Heller's operation) This has nothing to do with the heart. The upper part of the stomach is referred to as the cardia. The operation is consisting of cutting the spastic muscle in the lower esophagus and the upper stomach to relieve the spasm.

Esophageal function test: measuring pressures in various segments of the esophagus.

Exsanguination: a loss of huge amounts of blood.

Extra-bronchial: outside the bronchus.

345

Fascia: connective tissue located in various places throughout the body, and sometimes surrounds the muscles of a limb or divides them into compartments.

Femoral: relating to the femur or thigh e.g. the femoral artery located in the groin.

Fentanyl: a synthetic opioid narcotic analgesic stronger than morphine used to relieve strong pain or during anesthesia.

Fibroma: a benign tumor consisting mainly of fibrous tissue.

Fungate: to assume a fungal form or grow rapidly like a mushroom.

Gastroscopy: endoscopic examination of the stomach.

H. pylori: bacteria which claim to play a factor in duodenal ulcer formation.

Heller's Operation: Cutting the muscular layer of the lower esophagus.

Hemi-sternotomy: surgically cutting half of the sternum (breast bone) and retracting it in order to inspect the heart or the great blood vessels coming out of it.

Hemoptysis: coughing of blood.

Heparinization: to treat with heparin, an anticoagulant.

Hepatosplenomegaly: coincident enlargement of the liver and spleen.

Hilum: e.g., of the lung, is the part of the lung from which blood vessels enter the heart.

Histology: examining tissue under the microscope used mainly to discover presence or absence of malignancy.

Hodgkin's disease: a type of lymphoma present mainly in the mediastinum (the area between the two lungs surrounding the trachea and the main bronchi). Histologically it is characterized by the presence of large cells. It is less aggressive than other types of lymphoma present in different parts of the body and referred to as Non-Hodgkin's lymphomas.

Homocysteine: an amino acid formed by oxidation of homocysteine and excreted in the urine in homocystinuria. It also causes occlusion in the blood vessels similar to cholesterol but easier to treat by vitamins B, e.g. Folic acid, B6 and B12.

Hyperalimentation: the administration of nutrients by intravenous feeding which are of high caloric value.

Hypertrophy: increase in the size of an organ.

Hypovolemia: decrease in the volume of the circulating blood.

Ileus: obstruction of the bowel by lack of peristalsis e.g. after abdominal surgery.

Intercostal: situated or extending between the ribs.

Intima: the innermost layer of a blood vessel.

Intra-bronchial: situated or occurring inside the bronchial tubes.

Intubate: the introduction of a tube into the trachea to keep it open or restore patency if obstructed.

Iodoform: a yellow crystalline volatile compound used as an antiseptic dressing.

Ischemia: lack of blood supply and oxygen to tissues.

Jejunum: that part of the small intestine between the duodenum and the ileum.

Labyrinthitis: inflammation of the inner ear.

Laparoscopy: visual examination of the abdomen by means of a laparoscope.

Laparotomy: incision of the abdominal wall. Usually reserved for exploratory operations.

Lateral: relating to the side of the body.

Leukocytes: white blood cells.

Ligation: surgical process of tying up an anatomical channel, as a blood vessel.

Lobectomy: surgical removal of part of the lung (a lobe).

Lower GI: exam of the large bowel by x-ray of the colon.

Lumpectomy: excision of a breast tumor with a limited amount of surrounding tissue.

Lymphoma: malignant tumor of the lymph nodes.

Mandible: the lower jaw.

Manometer: a tool to measure pressure (e.g. in the esophagus).

Manubrium: upper part of the sternum (the breast bone).

Medial: close to the midline.

Mediastinal surgery: surgery in the area between the breast bone and the heart.

Mesentery: membranes connected with the bowels and carrying the blood vessels to it.

Metastasis: spread of a cancer from its initial or primary site to another part of the body.

MRI: Magnetic Resonance Imaging is recording of magnetic waves from different parts of the body and recording them on a receptive plate and does not entail the use of x-rays.

Mucosa: mucous membrane; e.g. the inner layer lining of the intestine.

Myoblastoma: a very rare tumor that is composed of cells resembling primitive myoblasts and is found inside the bronchus of a lung.

Myocardium: the middle muscular layer of the heart wall.

Narcosis: a state of stupor, unconsciousness caused by narcotics or other chemicals e.g., CO_2 narcosis is caused by retaining more CO_2 (carbon dioxide) than oxygen in the lung.

Necrosis: death of living tissue usually resulting from cutting its blood supply or severe infection.

Nitric Oxide: a result of adding an oxygen atom to Nitrous Oxide (N_2O to NO). It is produced in the body from the amino acids. Arginine and Citruline were found to protect the blood vessels from formation of plaques and have a vasodilatation effect. Its discoverer got the Nobel Prize.

Omentum: a fold of peritoneum extending from the stomach to the pubis like an apron.

Oximetry: measuring the percentage of oxygen in the circulation.

Papilloma: a mushroom-like tumor usually on the hands and fingers.

Para-aortic: surrounding the aorta e.g., para-aortic glands.

Paradoxically: unusual behavior e.g., paradoxical breathing which results from multiple rib fractures on both sides of the chest resulting in an abnormal type of breathing in which, instead of the chest expanding symmetrically, half of the chest rises then the other half. It necessitates treatment by a respirator till the fracture heals.

Parasitology: the study of diseases caused by parasites.

PARP: Poly ADP-ribose polymerase, an enzyme used by cancer cells to repair DNA damage.

Periappendicular abscess: abscess that surrounds the appendix resulting from its perforation caused by a delayed appendectomy in the case of appendicitis.

Pleural effusion: the presence of fluid in the pleural cavity.

Pneumonectomy: surgical removal of a whole lung.

Popliteal: of or relating to the back part of the leg behind the knee joint.

Prednisone: an anti-inflammatory drug stronger than cortisone and is used in the treatment of arthritis and severe spasm in the lung.

Prophylactically: precautions to prevent a disease from happening.

Proximal: a place before an object; opposite of distal; a place further from it.

Resection: surgical removal of all or part of an organ, tissue, or structure.

Retinaculum: a connective tissue band in the wrist thickened by repeated use of the wrist as in long periods of typing causing pressure on the median nerve.

Saphenous Vein: one of two chief superficial veins of the leg e.g., long Saphenous and short Saphenous. The short one is disposable and can be used for venous or coronary artery bypass.

Sarcomere: a unit of striated muscle.

Scalenus: any of three deeply situated muscles on each side of the neck and each extends from the transverse processes of two or more cervical vertebrae to the first or second rib.

Segmentectomies: segmental resection, e.g. part of a lobe in the lung.

Sentinel node: the first lymph node to receive lymphatic drainage from a primary tumor, e.g., breast cancer.

Shunt: a connection between two blood vessels.

Splenectomy: surgical removal of the spleen.

Splenomegaly: enlargement of the spleen.

Squamous cell: a cell derived from squamous epithelium.

Stenosis: a narrowing or constriction of a bodily passage or orifice e.g., trachea.

Sternum: the bone in front of the chest between the ribs. Its upper part is called the manubrium, the middle part is called the body and the lower part is called the xyphoid process.

Subclavian: below the clavicle, e.g., subclavian vessels which pass from the neck to the arm under the clavicle.

Sural Nerve: a sensory nerve in the calf which is disposable and can be used for bridging the gap between two severed segments of an important nerve.

Syncope: loss of consciousness resulting from insufficient blood flow to the brain.

Tamoxifen: an estrogen receptor modulator that acts as an estrogen antagonist in breast cancer when it is positive for estrogen receptors.

Tamponade: fluid around the heart which prevents it from expanding resulting in low pulse pressure (the difference between systolic and diastolic blood pressure).

Thoracotomy: opening the chest by cutting between the ribs.

Thrombose: the formation of a blood clot within a blood vessel.

Thrombectomy: removal of a thrombose from an artery.

Thrombokinase: a chemical enzyme that is supposed to dissolve blood clots in a blood vessel.

Thrombolysis: dissolving a thrombose or blood clot using chemicals e.g., Thrombokinase.

Thrombus: a clot of blood formed within a blood vessel.

Tibial: relating to the tibia which is the larger of the two bones of the leg as referred to as the shin bone.

Tracheo-bronchial: relating to the trachea and the main stem bronchi.

Upper GI: radiographic examination of the esophagus and stomach using a radio-opaque contrast e.g., barium.

Ureter: a delicate tube connecting the kidney to the urinary bladder which is small in diameter.

Vascularization: means introducing blood to an a-vascular area.

Venogram: radiography of a vein after injection of an opaque substance.

Vincristine: an anti-neoplastic agent.

WBC: white blood count, counting the percentage of white blood cells in the blood.

BIBLIOGRAPHY

GOOD MEDICAL PRACTICE

1. *Principle and Practice of Medicine* by Davidson. Published by Churchill Livingstone (2006)

2. *Successful Surgery* by R. Baker M.D., published by Simon and Schuster (1996)

3. *Tomorrow's Doctor* by B. Natelson M.D, published by Insight Books (1990)

4. *The Successful Physician* by M. Zaslove, published by Aspen Publication (1998)

5. *Working with Your Doctor* by N. Keene, published by O'Reilly & Associates (1998)

6. *Doctoring: The Nature of Primary Care Medicine* by Eric J. Cassell, published by Oxford University Press (1997)

CANCER

1. *The Cancer Survival Guide* by Peeley & Bashe, published by Broadway Books (2005)

2. *Options—The Alternative Cancer Therapy Book* by R. Walters, published by Avery Books (1993)

3. *How I Conquered Cancer Naturally* by Mae Eydie, published by Avery Publishing (1992)

4. *Breast Cancer—the Complete Guide* by Hirshaut M.D. & Pressman M.D., published by Bantam Books (1996)

5. *Cancer: Principles and Practice of Oncology—Lung Cancer* by Vincent T. DeVita Jr., et al, published by Lippincott, Williams & Wilkins (2006)

6. *Cancer: Principles and Practice of Oncology—Breast Cancer* by Vincent T. DeVita Jr., et al, published by Lippincott, Williams & Wilkins (2006)

7. *Choices—The Most Complete Sourcebook for Cancer Information, 4th edition* by Marion Morra & Eve Potts, published by Quill (2003)

8. *Everyone's Guide to Cancer Therapy* by Malin Dollinger MD, et al, revised 2nd edition, published by A Universal Press Syndicate Company (2008)

9. *Cancer Cure* by Gary L. Schine with Ellen Berlinsky, Ph D published by Kensington Books (1996)

10. *A Cancer Sourcebook for Nurses* published by American Cancer Society (2004)

11. *Breast Cancer* by Steve Austin ND et al, published by Prima (1994)

12. *Cancer Principles and Practice, 6th edition* by Vincent T. DeVita Jr. et al, published by Lippincott, Williams & Wilkins (2001)

13. *Non-Hodgkins Lymphoma* by Lorraine Johnston, published by O'Reilly & Associates (1999)

14. *Options* by Richard Walters, published by Avery (1992)

15. *The Breast Cancer Prevention Program* by Epstein, Steinman & LeVert, published by Macmillan USA (1997)

LOVE & COMPASSION

1. *Peace, Love & Healing* by Bernie S. Siegel, MD, published by Quill (2001)

2. *Loving Kindness: The Revolutionary Art of Happiness* by Sharon Salzberg, published by Shambhala Library (2004)

3. *How to Break Bad News: A Guide for Health Care Professionals* by Robert Buckman, MD, published by The Johns Hopkins University Press (1992)

4. *The Future of Love* by Daphne Rose Kingma, published by Doubleday (1998)

5. *Return to Love* by Marianne Williamson, published by Harper Collins (1992)

6. *Journey of Love* by Alan Mesher, published by Arnan Publishing (1982)

7. *The Courage to Give* by Jackie Waldman et al, published by Conari Press (2000)

MIND—BODY HEALING

1. *Timeless Healing* by Herbert Benson M.D., published by A Fireside Book, (1996)

2. *The Wellness Book* by Herbert Benson M.D, published by A Fireside Book (1992)

3. *Healing and the Mind* by Bill Moyers, published by Broadway Books (1979)

4. *Quantum Healing* by Deepak Chopra, M.D. published by Bantam Books (1989)

5. *Reinventing Medicine* by Larry Dossey, M.D. published by Harper San Francisco (1999)

6. *Ageless Body, Timeless Mind* by Deepak Chopra, M.D. published by Three Rivers Press (1993)

7. *Remarkable Recovery* by Hirshberg & Barasch, published by Riverhead Books (1996)

8. *Molecules of the Motion—The Science beyond Mind Body Medicine* by Candice Pert, published by Scribner (1997)

9. *Your Mind Can Heal You* by Frederick Bailes published by DeVorss Company (1971)

10. *Thoughts are Things* by Ernest Holmes, published by Health Comm. Inc. (1967)

11. *Body Mind Balancing* by OSHO, published by St. Martin's Griffin (2003)

12. *The Miracle of Mind Dynamics* by Joseph Murphy, DD, DRS, PhD, LLD, published by Reward Books (1964)

13. *The Healing Intelligence* by Harry Edwards, published by Westwood Publishing Co. (1971)

14. *The Hidden Meaning of Illness* by Bob Trowbridge, M.Div., published by A.R.E. Press (1997)

15. *Mind—Body Medicine* edited by Dr. Alan Watkins, published by Churchill Livingstone (1997)

16. *Mind—Body Medicine: How to Use Your Mind for Better Health* edited by Daniel Goleman, Ph.D. & Joel Gurin, published by Consumer Report Books (1996)

17. *Healing Beyond the Body* by Larry Dossey, MD, published by Shambhala (2003)

18. *Mingling Minds* by Ervin Seale, published by DeVorss Publications (1986)

19. *The Dynamic Laws of Healing* by Catherine Ponder, published by DeVorss (1966)

20. *Mind/Body Medicine* by Daniel Goleman and Joel Gurin, published by Consumer Reports Books (1996)

ACCIDENT, SYNCHRONICITY, DESTINY OR COINCIDENCE

1. *There Are No Accidents* by Robert Hopcke, published by Riverhead Books (1997)

2. *Small Miracles of Love and Friendship* by Halberstam & Leventhal, published by Adams Media Corp. (1999)

3. *Jung on Synchronicity and the Paranormal* By Roderick Main, published by Princeton University Press (1997)

4. *Synchronicity Through the Eyes of Science, Myth and the Trickster* by Allan Combs and Mark Holland, published by Marlowe and Company (1996)

ATTITUDE

1. *Love is Letting Go of Fear* by Jampolsky M.D., published by Celestial Arts (1979)

2. *Love is the Answer* by Jampolsky M.D., published by Bantam Books (1990)

3. *The 12 Principles in Attitudinal Healing* by Jampolsky M.D. published by Beyond Words Publishing (2000)

4. *Charming Your Way to the Top* by Michael Levine, published by the Lyons Press (2004)

5. *Anatomy of Hope* by Jerome Groopman, MD, published by Random House (2004)

6. *The Courage to Laugh* by Allen Klein, published by Penguin Putnam Inc. (1998)

7. *Learned Optimism: How to Change Your Mind and Your Life* by Martin E. P. Selegman, published by Vintage (2006)

8. *Hope for Tough Times* by Mary J. Nelson, published by Spire (2006)

9. *The Scalpel and the Soul Encounters with Surgery, the Supernatural, and the Healing Power of Hope* by Allan J. Hamilton, MD, FACS, published by Tarcher/Penguin (2008)

10. *Hope Again: When Life Hurts and Dreams Fade* by Charles R. Swindoll, published by Word Publishing (1996)

11. *Optimist: A Practical Guide to Achieving Happiness* by Lucy MacDonald, published by Chronicle Books (2004)

ALTERNATE MEDICINE

Mayo Clinic and Duke University Manuals on Alternative Healing Therapeutics. The accepted alternative medicines for me are:

1. *No more Heart Disease* by Luis Ignarro, M.D. (2006)

   1. Nitric Oxide in treatment of arteriosclerotic disease.

   2. Chelation Therapy.

   (Many books under this title are found on the Internet.)

   3. Antioxidant Therapy.

   Some doctors combined modern medicine with alternative therapies.

2. *Balanced Healing* by Larry Altshuler, published by Harvard Press (2004)

3. *A Total Nutrition* by Victor Herbert, published by Martins Griffin, New York. (1995)

4. *A Consumer Guide to Chelation Therapy* by Harold & Arline Brecker, published by Health Savers Press (1997)

5. *Chelation Therapy* by Martin Dayton, MD, DO, published by Martin Dayton (1995)

6. *The Alternative Medicine Handbook* by Barrie Cassileth, PhD, published by W.W. Norton & Co. (1998)

7. *Complementary & Alternative Medicine* by Donald W. Novey, MD, published by Mosby (2000)

8. *The Consumer Guide to Alternative Medicine in association with the American Association of Naturopathic Physicians*, published by Publications International, Ltd. (1997)

9. *Consciousness and Healing* by Marilyn Schlitz and Tina Amorok with Marc S. Micozzi, published by Elsevier Churchill Livingstone (2005)

10. *Alternative Healing*, by Arnold Fox, MD and Barry Fox, Ph.D., published Career Press (1996)

11. *Clinician's Guide to Holistic Medicine* by Robert A. Anderson, published by McGraw-Hill (2001)

12. *The Medical Advisor: The Complete Guide to Alternative and Conventional Treatments* by the editors of Time-Life Books, published by Time-Life Books (2000)

13. *The Coenzyme Q10 Phenomenon: The Breakthrough Nutrient that Helps Combat Heart Disease, Cancer, Aging and More* by Stephen T. Sinatra, MD, FACC, published by Keats Publishing (1998)

14. *The Sinatra Solution: Metabolic Cardiology* by Stephen T. Sinatra, MD, FACC, published by Basic Health Publications (2005)

15. *Alternative Medicine: The Definitive Guide* by The Burton Goldberg Group, published by Future Medicine Publishing (1995)

16. *New Medicine Complete Family Health Guide* by Professor David Peters et al, published by DK Publishing (2005)

ANTI-AGING MEDICINE

1. *You Staying Young: The Owner's Manual for Extending Your Warranty* By Michael F. Roizen, MD, et al, published by Simon & Schuster (2007)

2. *Age Smart: Discovering the Fountain of Youth at Midlife and Beyond* by Jeffrey Rosensweig, Ph. D. & Betty Liu published by Pearson Prentice Hall (2006)

3. *Grow Young with HGH* by Dr. Ronald Klatz published by Harper Collins Publishers (1997)

4. *Growth Hormone Reversing Human Aging Naturally: The Methuselah Factor* by James Jamieson et al, published by Safegoods and LNN (1997)

5. *Ageless: The Naked Truth about Bioidentical Hormones* by Suzanne Somers, published by Crown (2006)

6. *Ageless Body, Timeless Mind: The Quantum Alternative to Growing Old* by Deepak Chopra, MD, published by Harmony Books (1993)

7. *Resetting the Clock* by Elmer Cranton, MD, et al, published by M. Evans & Company (1996)

8. *The Longevity Bible Eight Essential Strategies for Keeping Your Mind Sharp and Your Body Young* by Gary Small, MD, published by Hyperion (2006)

ON MIRACLES

1. *Gifts from a Course in Miracles*, edited by Frances Baughan, published by Putnam Publishing (1995)

2. *The Message of a Course in Miracles* by Kenneth Wapnick, published by Foundation of the Course in Miracles (2003)

3. *Where Miracles Happen* by Joan Wester Anderson, published by Random House (1994)

4. *With God All Things Are Possible*, published by Bantam Books (1986)

5. *How to Make the Impossible Possible* by Robert Anthony, M.D. published by Berkley Books (1996)

6. *Small Miracles of Love and Friendship* by Halberstam & Leventhal, published by Adams Media Corp. (1999).

7. *Let Go, Let Miracles Happen* by Kathy Cordova, published by Conari Press (2003)

8. *Practical Miracles* by John Gray, PhD, published by Harper Collins (2008)

9. *Remarkable Recovery* by Caryle Hershberg & Marc I. Barasch, published by Riverhead Books (1995)

10. *It's a Miracle 1 & 2* by Richard Thomas, published by Bantam Dell (2003)

11. *Journey of Love—A Formula for Mastery & Miracles* by Alan Mesher, published by Quartus Books (1987)

12. *Miracles Made Easy* by Jack Groverland, published by Miracle Publishing Co. (1985)

13. *A Course in Miracles*, published by the Foundation for Inner Peace (1992)

14. *Miracles are Still Happening* by A. L. & Joyce Gill, published by Whitaker House (1982)

15. *You Can Work Your Own Miracles* by Napoleon Hill, published by Random House Publishing Group (1971)

16. *Shortcut to a Miracle How to Change Your Consciousness and Transform your Life* by Michael C. Rann & Elizabeth Rann Arrott, published by Jeffers Press (2005)

INTUITION AND INSPIRATION

1. *Guide to Intuitive Healing* by Doctor Judith Orloff, published by Three Rivers Press (2000)

2. *Surgical Intuition* by Abernathy Hamm, published by Hanley Belfus (1995)

3. *Inspiration* by Dr. Wayne Dyer, published by Hay House (2006)

4. **Edgar Cayce** from *Encyclopedia of Healing* by Reba Karp, published by Warner Books (1986)

5. *The Story of Edgar Cayce –There is A River* by Thomas Sugrue, published by A.R.E. Press (1973)

6. *Natural-Born Intuition* by Lauren Thebodeau, PhD, published by Career Press (2005)

7. *Living Inspired* by Akiva Tutz, published by Targum Press (1993)

8. *Health Intuition* by Karen Grace Kassy, Published by Hazelden (2000)

9. *Surgical Decision Making* by Eiseman and Wotkyns, published by Saunders (1978)

10. *Breaking the Rules* by Kurt Wright, published by CPM Publishing (1998)

11. *The Intuitive Healer* by Marcia Emery Ph.D., published by St. Martin's Griffin (1999)

12. *Intuitive Living* by Alan Seale, published by Weiser Books (2001)

13. *Divine Intuition* by Lynn Robinson, M.Ed., published by A Dorling Kindersley Book (2001)

14. *Discover Your Inner Wisdom Using Intuition, Logic and Common Sense to Make Your Best Choices* by Char Margolis, published by Fireside (2008)

PRAYER, AFFIRMATIONS AND MEDITATION IN MEDICINE

1. *Prayer is Good Medicine* by Larry Dossey, M.D. published by Harper San Francisco (1996)
2. *Healing Through Prayer* by Dossey and others, published by Anglican Book Centre (1999)
3. *Prayer, Fate and Healing* by Cain & Kaufman, published by Rodale (1999)
4. *Faith Factor* by Dale Mathews M.D. by Penguin Books (1998)
5. *Meditation as Medicine* by Doctor Khalsa M.D. & Cameron Stauth, published by Pocket Books (2001)
6. *Science and Health* by Mary Baker Eddy, published by The First Church of Christ (1934)
7. *The Anatomy of Healing Prayer* by Ernest Holmes, published by DeVorss (1991)
8. *The Handbook of Positive Prayer* by Hypatia Hasbrouck, published by Unity Books (1995)
9. *Prayer* by Ernest Holmes, published by Tarcher/Penguin (2008)
10. *Words That Heal* by Douglas Bloch, published by Bantam Books (1971)
11. *Meditations to Heal Your Life* by Louise L. Hay, published by Hay House (1994)

CREATIVITY IN MEDICINE

1. *Creativity* by Mathew Fox, published by Putnam (2002)
2. *The Soul of Creativity* by Tona Pearce Myers, published by New World Library (1999)
3. *Creating Minds* by Howard Gardner, published by Basic Books, (1993)
4. *The Creativity Book* by Eric Maisel, published by Putnam, (2000)
5. *Creative Visualization* by Melita Denning & Osborne Phillips, published by Llewellyn Publications (1987)
6. *Creative Visualization* by Shakti Gawain, published by World Library, (1995)
7. *Creative Meditation* by Richard Peterson, published by A.R.E. Press (1990)

8. *Creativity* by Mihaly Csikszentmihalyi, published by Harper Collins Publishers (1996)

9. *Creative Mind & Success* by Ernest Holmes, published by Tarcher/Penguin (1919)

10. *Creativity—Unleashing the Forces Within* by OSHO, published by St. Martin's Griffin (1999)

11. *Breakthrough to Creativity* by Shafica Karagulla MD, published by DeVorss (1972)

12. *Creative Visualization* by Shari L. Just PhD D et al, published by Alpha (2002)

13. *Out of the Box* by Rob Eastaway, published by Duncan Baird Publishers (2007)

14. *The Artists Way: A Spiritual Path to Higher Creativity*, by Julia Cameron, published by Tarcher/Putnam (2002)

15. *Secret, The Power, Chapter on Feeling is Creation* by Rhonda Byrne, published by Atria Books (2010), pages 83-92.

16. *The Vibrational Universe: Harnessing the Power of Thought to Consciously Create Your Life* by Kenneth James Michael MacLean, published by Loving Healing Press (2006)

17. *Unlock Your Mind: A Practical Guide to Deliberate and Systemic Innovation* by Dennis Sherwood, published by Gower (1998)

18. *Cracking Creativity: The Secrets of Creative Genius* by Michael Michalko, published by Ten Speed Press (2001)

MODERN MEDICAL INVENTORS: 1950 AND FORWARD

1. *The Genius of C. Walton Lillehei & The History of Open Heart Surgery* by Daniel A Goor MD, published by Vantage (2007)

2. *Journey into the Heart—A Tale of Pioneering Doctors* by David Monagan, published by Gotham Books (2007)

3. *The Rebuilt Man—The Miracle of Transplant Surgery* by Fred Warshofsky, published by Crowell (1965)

4. *Partners of the Heart—Vivien Thomas and his work with Alfred Blalock* by Vivien Thomas, published by Penn (1998)

5. *Every Second Counts—The Race to Transplant the First Human Heart* by Donald McRae, published by Putnam (2007)

6. *Cooley* by Harry Minitree, published by Harper's Magazine Press (1973)

7. *The Medical Discoveries of Edward Bach, Physician* by Nora Weeks, published by Keats (1973)

8. *Alton Oschsner—Surgeon of the South* by John Wilds & Ira Harkey, published by Louisiana State University Press (1990)

9. *Amazing Adventures of a Heart Surgeon* by Domingo Liotta MD, published by iUniverse, Inc. (2007)

10. *Bypassing Bypass Surgery* by Elmer M. Cranton, MD, published by Hampton Roads (2001)

11. *The Miracle Finders* by Donald Robinson, published by Mackay (1976), Chapters 6 & 7.

    a) Forssman for catheter insertion
    b) Cournand for the 1st cardiac catheterization
    c) Blalock & Taussig for their work with blue babies
    d) Gross for ligating patent ductus
    e) Hufnagel for aortic valve in descending aorta
    f) Gibbon for inventing the heart-lung machine
    g) Albert Starr for valve replacement
    h) Mason Sones for coronary angiography
    i) Lillehei for the first open heart surgery
    j) DeBakey for the roller pump and pioneering carotid surgery
    k) Cooley for inserting the first artificial heart
    l) Shumway for inventing the technique for heart transplantation
    m) Bernard for doing the first heart transplant
    n) Favoloro for the first right coronary bypass in May, 1967
    o) Dudley Johnson for the first left coronary bypass one year later

HUMILITY

1. *Humility: True Greatness* by C. J. Mahaney, published by Multnomah Books (2005)

SERENDIPITY

1. *Serendipity—Accidental Discoveries in Science* by Royston Roberts, published by John Whaley and Sons, (1989)
2. *Happy Accidents: Serendipity in Modern Medical Breakthroughs* by Morton A. Meyers, MD, published by Arcade Publishing (2007)

SERENITY

1. *Serenity* by Jane Nelsen, published by Conari Press (2008)
2. *Secrets of Serenity: Timeless Wisdom to Soothe the Soul*, published by Courage Books (2004)

HEALING BY ENERGY POWER

1. *Instant Healing* by Serge Kahili King, PhD, published by Renaissance Books (2000)
2. *Bursting with Energy* by Frank Shallenberger, MD, published by InMED (2002)
3. *Energy Medicine* by Donna Eden, published by Tarcher/Putnam (1998)
4. *Energy Forever* by Sid Kirchheimer & Gale Maleskey, published by Dell Publishing (1997)

HEALING BY MENTAL ENERGY

1. *Instant Healing* by Serge Kahili King, PhD, published by Renaissance Books (2000)

FAITH HEALING

1. *Christ Enthroned in Man* by Cora D. Fillmore, published by Unity Books (1981)
2. *Why Faith Matters* by David J. Wolpe, published by Harper One (2008)

HEALING BY THE TOUCH OF HANDS

1. *A Touch of Hope* by Dean Kraft, published by Berkley Books (1998)
2. *Portrait of a Psychic Healer* by Dean Kraft, published by Putnam (1981)
3. *The Secret of Powerful Healing* by Uri Geller, published by Element Books (1999)

   The above two authors are endowed with unusual magnetic energies.

4. *A Gift for Healing—How you can use Therapeutic Touch* by Deborah Cowens, MSN, RN, ANP, published by Crown Trade Paperbacks (1996)

SPIRITUAL HEALING

1. *The Art of Spiritual Healing* by Joel Goldsmith, published by Harper San Francisco (1992)
2. *Clinician's Guide to Spirituality* by MacDougall & White published by McGraw-Hill (2001)
3. *The Art of Spiritual Healing* by Sherwood, published by Llewellyn Publications (2000)
4. *Awakening the Mystique Gift* by Jane Doherty, published by Hummel & Solvarr (2005)
5. *The Healing Connection* by Harold Koenig M.D., published by Word Publishing (2000)
6. *Healing Spiritually* by Christian Science Publishing Company, (1996)
7. *Prayer That Heals* by Ann Beals, published by The Bookmark, (2002)
8. *Spirit Body Healing* by Michael Samuels, MD, published by Wiley (2000)
9. *The Secret of Healing* by Jack E. Addington, published by Science of Mind Publications (1979)
10. *The Amazing Laws of Cosmic Mind Power* by Joseph Murphy, published by Parker Publishing Co. (1965)
11. *The Spiritual Dimension* by Ann Beals, published by The Bookmark (2007)

12. *Unity* by Eric Butterworth, published by Unity Books (1985)

13. *Point of Power* by Paul Hasselbeck, published by Prosperity Publishing House (2007)

14. *Jesus: Teacher & Healer* published by White Eagle Teaching & Publishing (1985)

15. *What God is Like* by James Dillet Freeman, published by Unity Books (1973)

16. *The Healing Power Within* by Ann Wigmore, published by Avery Publishers (1983)

17. *Awaken the Healer Within* by Rich Work & Ann M. Groth, published by Asini Publisher (1993)

18. *You Are the Answer* by Michael J. Tamura, published by Llewellyn Publisher (2007)

19. *What is the Father Like* by W. Phillip Keller, published by Bethany House Publisher (1996)

20. *Science and Health* by Mary Baker Eddy (1971)

21. *Healers on Healing* by Carlson and Shield, published by Jeremy P. Tarcher, Inc. (1989)

22. *The Extraordinary Healing Power of Ordinary Things* by Larry Dossey, MD published by Three Rivers Press (2006)

23. *The Spirit Body Healing* by Michael Samuels, MD & Mary Lane RN, PhD, published by John Wiley & Sons (2001)

24. *Invisible Acts of Power* by Caroline Myss, published by Free Press (2004)

25. *Christ the Healer* by F.F. Bosworth, published by Chosen Books (2008)

26. *Spiritual Healing for Today* by Raymond Charles Barker, published by DeVorss (1988)

27. *Spiritual Healing* by Jack Angelo, published by Element (1991)

IN PRAISE OF MEDICINE

1. *Medical Marvels-the Hundred Greatest Advances in Medicine* by Eugene Straus M.D. & Alex Straus, published by Prometheus Books (2006)
2. *The Medicine-From Hippocrates to Gene Therapy* by Paul Strathern, published by Carroll & Graf Publishers (2005)
3. *Essays of Denton A. Cooley-Reflections and Observations*, collected by Marianne Kneipp, published by Eakin Press (1984)

BOOKS ABOUT MALPRACTICE

1. *The Malpractice Epidemic* by Bernard Leo Remakus, published by Authors Choice Press (1990)
2. *Justice Overruled* by Judge Burton S. Katz, published by Warner Books (1997)
3. *The Myth of Moral Justice* by Thane Rosenbaum, published by Perennial (2004)
4. *The Malpractice Cure* by Edward D. McCarthy, Esq., published by Kaplan Publishing (2009)
5. *Designing Care* by Richard M.J. Bohmer, published by Harvard Business Press (2009)
6. *The Medical Malpractice Survival Handbook*, published by Mosby Elsevier (2007)
7. *Medical Malpractice on Trial* by Paul C. Weiler, published by Harvard University Press (1991)
8. *Legal Medicine 7th Edition* by S. Sandy Sanbar, MD, PhD, JD, FCLM, et al, published by Mosby Elsevier (2007)
9. *Capping Non-Economic Awards in Medical Malpractice Trials* by Nicholas M. Pace et al, published by Rand Corporation (2004)
10. *The Biggest Legal Mistakes Physicians Make and How to Avoid Them*, published by SEAK, Inc. (2005)
11. *Confidence-How to Succeed at Being Yourself* by Alan Loy McGinnis, published by Augsburg Publishing House (1987)
12. *Medical Malpractice Law & Litigation* by Beth Walston-Dunham, published by Thomson Delmar Learning (2006)

BOOKS THAT CRITICIZE THE MEDICAL PROFESSION

1. *A Wall of Silence* by Gibson & Singh, published by Lifeline Press (2003)
2. *Physician-Medicine and the Battle for Human Freedom* by Richard Leviton, published by Hampton Roads Publishing (2000)
3. *You—the Smart Patient* by Roizen M.D. & Oz M.D., published by Free Press (2006)
4. *Hands of Life* by Julie Martz, published by Bantam Books, (1998)
5. *Just Here Trying to Save a Few Lives—Tales of Life and Death from the ER* by Pamela Grim M.D., published by Warner Books (2000)
6. *Who Killed Health Care* by Regina Herzlinger, published by McGraw-Hill (2007)

WRONG DIAGNOSIS

1. *Internal Bleeding* by Wachter M.D. & Shojamia M.D., published by Ruggud Land (2004)
2. *How Doctors Think* by Jerome Groopman, published by Houghton Mifflin (2007)
3. *Complications* by Atul Gawande, published by Metropolitan (2002)
4. *Every Patient Tells a Story* by Lisa Sanders, MD, published by Broadway Books, (2009)
5. *Second Opinion* by Jerome Groopman, published by Penguin Books (2001)

BOOKS BY OR ABOUT DR. COOLEY

1. *Techniques in Cardiac Surgery* by Denton A. Cooley, M.D., published by Saunders (1984)

2. *Techniques in Vascular Surgery* by Denton A. Cooley, M.D. & Don C. Wukasch, published by Saunders (1979)

3. *Hearts by Thomas Thompson*, published by McCall Publishing Society (1973)

4. *Cooley* by Harry Minitree, published by Harper Magazine Press (1973)

5. *Heart Owners Handbook* by Texas Heart Institute, published by Wiley (1996)

6. *Surgical Treatment of Aortic Aneurysms* by Denton A. Cooley, M.D., published by Saunders (1986)

7. *Surgical Treatment of Congenital Heart Disease, 3rd edition* by Denton A. Cooley, M.D., et al, published by Lea & Febiger

8. *100,000 Hearts: A Surgeons Memoir* by Denton A. Cooley, M.D., published by Dolph Briscoe Center American History

The End

www.ingramcontent.com/pod-product-compliance
Lightning Source LLC
Chambersburg PA
CBHW031816170526
45157CB00001B/78